T0255371

Materials for Energy Storage, Generation and Transport

MATERIALS RESEARCH SOCIETY
SYMPOSIUM PROCEEDINGS VOLUME 730

Materials for Energy Storage, Generation and Transport

Symposium held April 2–4, 2002, San Francisco, California, U.S.A.

EDITORS:

Ricardo B. Schwarz
Los Alamos National Laboratory
Los Alamos, New Mexico, U.S.A.

Gerbrand Ceder
Massachusetts Institute of Technology
Cambridge, Massachusetts, U.S.A.

Steven A. Ringel
The Ohio State University
Columbus, Ohio, U.S.A.

Materials Research Society
Warrendale, Pennsylvania

CAMBRIDGE UNIVERSITY PRESS
Cambridge, New York, Melbourne, Madrid, Cape Town,
Singapore, São Paulo, Delhi, Mexico City

Cambridge University Press
32 Avenue of the Americas, New York NY 10013-2473, USA

Published in the United States of America by Cambridge University Press, New York

www.cambridge.org
Information on this title: www.cambridge.org/9781107411852

Materials Research Society
506 Keystone Drive, Warrendale, PA 15086
http://www.mrs.org

First published 2002
First paperback edition 2013

Single article reprints from this publication are available through
University Microfilms Inc., 300 North Zeeb Road, Ann Arbor, MI 48106

CODEN: MRSPDH

ISBN 978-1-107-41185-2 Paperback

CONTENTS

MATERIALS FOR SOLAR ENERGY

SOLID OXIDE FUEL CELLS

POSTER SESSION

MATERIALS FOR POWER
IN SPACE

DISORDERED AND NANOSCALE MATERIALS
FOR ENERGY APPLICATIONS

*Invited Paper

THERMOELECTRICS

PREFACE

Symposium V, "Materials for Energy Storage, Generation and Transport," was held April 2–4 at the 2002 MRS Spring Meeting in San Francisco, California. 66 papers were presented in eleven sessions, including one poster session. The sessions were well attended and the discussions were lively.

The symposium focused on materials issues in the low-scale generation, storage and transport of energy. The topic of the symposium was motivated by our need to address energy generation and distribution of power in today's society. It is clear that large-scale power generation will include debates on long-term societal issues such as safety and environmental impact, but enhancing the efficiency of day-to-day devices (transport, heating/refrigeration, home electronics, etc.) requires mainly the development of new materials with improved properties. This MRS symposium addressed the latter issues, focusing on photovoltaics, batteries, fuel cells and other small-scale devices. Each session opened with an invited talk giving the attendants an overview of the materials limitations of these various techniques.

Symposium organizers were: R.B. Schwarz, G. Ceder, and S.A. Ringel. Session chairs also included G.A. Nazri, D. Flood, M. Krumpelt, and L. Kazmerski.

The symposium organizers and the Materials Research Society gratefully acknowledge the financial support by the Los Alamos National Laboratory for this symposium. We also thank Steven C. Moss for help in the editing of the proceedings.

Ricardo B. Schwarz
Gerbrand Ceder
Steven A. Ringel

October 2002

MATERIALS RESEARCH SOCIETY SYMPOSIUM PROCEEDINGS

MATERIALS RESEARCH SOCIETY SYMPOSIUM PROCEEDINGS

Prior Materials Research Society Symposium Proceedings available by contacting Materials Research Society

Lithium Batteries

Mat. Res. Soc. Symp. Proc. Vol. 730 © 2002 Materials Research Society

Behavior of LiMn$_2$O$_4$ Single Crystals as Battery Cathodes

María Ángeles Monge[1], José Manuel Amarilla[1], Enrique Gutiérrez-Puebla[1], Juan Antonio Campa[2], and Isidoro Rasines[1]
[1]Instituto de Ciencia de Materiales de Madrid, CSIC, Cantoblanco, 28049 Madrid, Spain.
[2]Facultad de Ciencias Geológicas, UCM, 28040 Madrid, Spain.

ABSTRACT

This paper deals with: i) the growth of LiMn$_2$O$_4$ crystals by electrocrystallisation; ii) the response of these crystals in lithium cells; and iii) a method to follow their structural and morphological changes while working as electrodes. The defects present in LiMn$_2$O$_4$ crystals are determined after refining the occupation at every site of the spinel structure by single-crystal X-ray diffraction analyses, which lead to define two possible paths for the process of Li deinsertion-insertion. It is shown how the crystals studied follow one of these paths during cycling of the battery by virtue of a dynamic mechanism consisting in Mn migrations cooperatively induced by Li insertion and extraction.

INTRODUCTION

Until recently the electrochemical behavior of LiMn$_2$O$_4$ has been always studied on powdered samples. LiMn$_2$O$_4$ crystals large enough to solve their crystal structure by single-crystal X-ray diffraction (XRD) were not grown until recently [1]. LiMn$_2$O$_4$, like the mineral spinel, crystallises in the cubic system, space group (S.G.) $Fd\overline{3}m$, $Z = 8$, with Li at tetrahedral $8a$ sites, Mn at the octahedral $16d$ positions, oxygen at $32e$ (u,u,u), with $a = 8.2483(6)$ Å and $u = 0.26320(18)$ [1] taking the origin at the inversion center, $\overline{3}m$. More recently, superlattice cells have also been described [2] for crystals which were mixed with powdered LiMn$_2$O$_4$ and electrochemically delithiated. After growing LiMn$_2$O$_4$ crystals by various techniques, electrocrystallisation led the authors to obtain good-quality black single crystals of various sizes from 0.1 to 1 mm, by systematically varying growth conditions such as flux, temperature, time, voltage and current. All these crystals were examined by single-crystal XRD.

EXPERIMENTAL DETAILS

LiMn$_2$O$_4$ crystals were grown in a three-electrode cell (Pt electrodes 1.5 mm diameter at a 10 mm distance, rotary Pt crucibles and a Pt wire 0.3 mm diameter as reference electrode) from mixtures of analytical grade reagents which were heated in an open-crucible furnace [3]. Once melted the mixtures were allowed to equilibrate for one hour and a half after reaching the operating crucible temperature. LiBO$_2$ (5 g), Mn$_2$O$_3$ (3 g) and Na$_2$MoO$_4$ (30 g), a temperature of 870 °C which was maintained for 2 hours, a potential of 0.30 V and currents of 150-175 mA were employed for batch **A**; and LiBO$_2$ (5 g), Mn$_2$O$_3$ (2 g), and Na$_2$MoO$_4$ (40 g), a temperature of 960 °C for 12 hours, 0.15 V and 19-26 mA for batch **B**. Crystals adhered to the electrode were separated from the matrix and washed in warm water. One crystal as grown from each batch was carefully selected: **A1** was the most representative crystal of the majority in cell parameter, occupations, and composition; and **B1**, the most different from **A1**. After confirming by DSC, differential scanning calorimetry, from −70 to +90 °C, that neither **A1** nor **B1** showed the phase transition reversibly associated to some polycrystalline samples [4], both were employed as cathodes for the electrochemical tests.

The Li extraction/insertion was performed at room temperature by galvanostatic cycling in Li metal cells using $LiPF_6$ (ethylene carbonate + dimethyl carbonate) as electrolyte. About 15 mg of **A1** or **B1** crystals were mixed with 16 mg of *Super P ™* carbon black from *MMM Carbon* in an orbital shaker and placed into a two-electrode *Swagelock ™* cell inside an argon glove box. The cells were cycled between 3.0 and 4.9 V at room temperature using a very small constant-current density, $j = 30$ μAcm^{-2}, in order to favor the homogeneous diffusion of the Li ions through the crystals. After cycling, the positive electrode was washed with propylene carbonate and methane trichloro and dried in air. Because they had large sizes, the crystals were easily separated. $Li_xMn_2O_4$ crystals were obtained from **A1** and **B1** at selected steps of the cycling: **A2** and **B2**, after first charge at 4.4 and 4.9 V, respectively; **A3** and **B3**, after first discharge at 3.3 V; and **B4**, after ten cycles between 3.0 and 4.9 V finishing in charge. The Li content, **x**, was determined from the amount of charge drained assuming 100 % Coulomb efficiency.

More than twenty crystals were analysed as indicated in [5].To study possible superstructures [4], the reciprocal space was carefully explored at various temperatures (20, 10, 0, -20, -50, -70 °C). Final data were collected at room temperature. With the purpose of having high redundancy in the spectrum, a hemisphere of the reciprocal space was collected for each crystal. The 2θ range was 4-31°, the number of the collected reflections varying between 803 and 845 (55-59 independent). Additional details can be found in [5]. All the crystal structures were solved refining the occupation in every site of the spinel structure.

RESULTS

The as grown crystals had octahedral habit like **A1** of figure 1. Figure 2 shows the first charge-discharge curves of **A1** and **B1** crystals during the first cycle of charge and discharge. The charge capacity of the cell assembled using ground **B1** crystals was close to that of the **B1** single crystal, only about 10 % higher. This result indicates that the crystals responded as fine-particle powder. For both **A1** and **B1**, the Li extracted during the first charge is larger than that inserted on discharge.

Figure 1. $LiMn_2O_4$ octahedral **A1** crystals showing 0.2 mm edges.

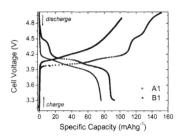

Figure 2. First galvanostatic charge/discharge curves ($j = 30$ μAcm^{-2}) of **A1** and **B1** crystals.

The cycling behavior of $LiMn_2O_4$ single crystals in a Li-cell is presented here. Figure 3 represents the curves recorded during cycling **B1** crystals between 3 and 4.9 V. It shows the high capacity retention that the studied crystal exhibits, since the discharge capacity only decreases from 78.2 to 67.4 mAhg^{-1} after ten cycles even at the small constant-current density used, which evidences a high reversibility. The photograph of figure 4 is the result of the first study by scanning electron microscopy of the morphological changes operated in a $LiMn_2O_4$ crystal after working as electrode in a Li cell during several cycles.

From all the crystals studied by XRD, an exciting and recurring result was obtained: the tetrahedral *8a* positions are shared by Li and at least 0.06 of Mn; and the population factors for the oxygen atom at *32e* always refine to 100 % occupancy. The structural data as well as the compositions obtained after solving the crystal structures of **A1-A3** and **B1-B4** single crystals are listed in tables I and II respectively. The analysis of these tables informs on the Li intercalation and the structural changes that it provokes.

DISCUSSION

The charge/discharge curves of figure 2 show two well-defined plateaus centred at ca. 4 V, which indicates that the lithium extraction/insertion takes place in two main steps as for powdered samples [6]. For $E \geq 4.4$ V, the **A1** curve shows another plateau which slightly polycrystalline nonstoichiometric samples give too [7,8]. In the case of **B1** the absence of any redox process in that voltage region indicates that **B1** contains more defects than **A1**, as single crystal XRD analyses confirm.

In the case of **A2**, the sample from batch **A** after charge, the cell parameter (table I) decreases as could be expected due to the Li extracted and the Mn oxidation; a peak of about 1.5 e$^-$ appears at *8b* positions in the Fourier difference; and simultaneously a decreasing in the *16d* occupation takes place. As *16d* multiplicity is twice that of *8b*, it seems quite plausible to think that the amount of Mn that disappears from the octahedral *16d* positions migrates evenly to *8b* and to *8a*.

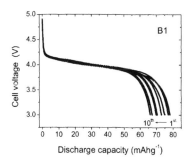

Figure 3. Discharge curves recorded during ten charge-discharge cycles in a Li cell.

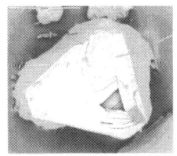

Figure 4. B4 crystals after working during cycling of **B1** crystals at $j = 30$ μAcm^{-2}.

Table I. Single-crystal structural data[1] and compositions of $Li_xMn_2O_4$ spinels from batch **A**

Crystal	A1	A2	A3
a/Å	8.236(1)	8.096(2)	8.213(2)
u	0.2632(5)	0.262(1)	0.2624(1)
R	0.03	0.07	0.07
Mn at *16d*	0.99(2)	0.94(2)	0.98(2)
Mn at *8a*	0.06(2)	0.12(4)	0.08(2)
Li at *8a*	0.95(2)	0.20(4)	0.65(2)
Mn at *8b*	-	0.06(2)	0.02(1)
x	-	0.19	0.63
Composition	$Li_{0.95}Mn_{2.04}O_4$	$Li_{0.20}Mn_{2.06}O_4$	$Li_{0.65}Mn_{2.06}O_4$

[1] S.G. $Fd\bar{3}m$ (No. 227), $Z = 8$. *a*, unit-cell parameter; *u*, oxygen positional parameter; R, discrepancy factor between determined and calculated XRD data. **x**, Li content as established during electrochemical cycling from deintercalation-intercalation data.

A2 transforms into A3 after the first discharge, showing an increase in the occupation of *16d* accompanied of a decrease at *8a* and *8b* (table I). This seems to indicate that the structural changes associated to the Li deintercalation are practically reversible. The remaining Mn at *8a* hinders the total reinsertion of Li. This accounts for the lower capacity in discharge than in charge during the first cycle, as had been commonly observed but not explained in the case of polycrystalline samples.

As to the **B1** crystals, all those analysed (table II) show some defects: i) about 10% of vacant positions at *16d*, ii) around 0.14 Mn sharing with Li the tetrahedral *8a* sites, and iii) a peak of near 1e⁻ at *8b* which was assigned to a little Mn, since otherwise it would imply such an amount of Li that would not match with the number of vacants at *16d* and both positions are too close to

Table II. Single-crystal structural data[1] and compositions of $Li_xMn_2O_4$ spinels from batch **B**

Crystal	B1	B2	B3	B4
a/Å	8.258(2)	8.097(1)	8.237(3)	8.151(5)
u	0.2633(6)	0.2625(5)	0.2627(6)	0.2627(5)
R	0.040	0.039	0.053	0.026
Mn at *16d*	0.91(3)	0.87(3)	0.89(3)	0.88(2)
Mn at *8a*	0.14(2)	0.12(2)	0.13(2)	0.10(1)
Li at *8a*	0.86(2)	0.10(2)	0.60(2)	0.10(1)
Mn at *8b*	0.03(1)	0.04(1)	0.06(1)	0.03(1)
Mn at *16c*	-	0.04(1)	-	0.06(1)
x	-	0.10	0.60	0.10
Composition	$Li_{0.86}Mn_{1.99}O_4$	$Li_{0.10}Mn_{1.98}O_4$	$Li_{0.60}Mn_{1.97}O_4$	$Li_{0.10}Mn_{2.01}O_4$

[1] Symbols, as in table 1.

be simultaneously occupied. Furthermore, distances M-O in this tetrahedral site, 1.600(8) A, are short, but quite similar to those for some Mn atoms in Mn_2O_3 and $MnO(OH)$ [9,10]. Can it be that Mn atoms at tetrahedral positions in **B1** are responsible for the high value of its cell parameter.

After first charge, **B2** denounces a decrease of the $8a$ and $16d$ occupations (table II), as well as a small increase at $8b$ and a new peak appearing at $16c$, which was also assigned to a low occupation of Mn. The total amount of Mn remains constant but a migration mainly from $16d$ to $16c$ positions is clearly observed.

In sample **B3** after the first cycle, a light increase of the occupation at both $16d$ and $8b$ and the disappearance of Mn at $16c$ are observed (table II), meaning this a nearly complete return of Mn to initial positions. The occupation of $8a$ also increases due to the Li intercalation. No changes were observed either in the oxygen atom or in the total amount of Mn. These observations seem to confirm that the structural changes associated to the Li deintercalation are practically reversible. The Mn occupations at both $8a$ and $8b$ shared positions diminish after cycling, while the electronic density at $16c$, about $2e^-$, doubles that of **B2** after first charge. This increase could be responsible for the progressive loss of capacity described for $LiMn_2O_4$.

All the crystals studied by single crystal XRD adjust to two models whose crystallographic formulas can be written: $[Li_{1-x}Mn_x]_{8a}[Mn_2]_{16d}O_4 \Leftrightarrow [Li_{1-x-}Mn_xMn_y]_{8a}[Mn_{2-y}]_{16d}[Mn_y]_{8b}O_4$, and $[Li_{1-x}Mn_x]_{8a}[Mn_{2-x-y}]_{16d}[Mn_y]_{8b}O_4 \Leftrightarrow [Li_{1-x-}Mn_x]_{8a}[Mn_{2-x-y-z}]_{16d}[Mn_y]_{8b}[Mn_z]_{16c}O_4$, for batches **A** and **B** respectively (Figure 5). Most of the studied crystals adapt to the first model, and only those that have been long time at higher temperature during electrocrystallisation show the degree of defects of the second.

Three actual possibilities established from experimental data are represented in the scheme of figure 5. Taking into account that the $8a$ position allows up to about 14% Mn sharing the position with Li (table II), the amount of Mn at $8b$ is always given by the difference between this upper limit, 14 %, and the amount of Mn coming from synthesis in the starting material. Once these percentages are surpassed either by the process of charge (**A2**), or by prolonged electrocrystallisation (**B1**), the Mn migration continues during charge to the octahedral $16c$ positions (**B2**). All these migrations are reversible in the first cycle, in such a way that **B3** and **A3** practically coincide with **B1** and **A1** respectively. As the cycling advances, some Mn remains at $16c$ (**B4**). The possibility of some Li sharing any of the indicated positions is not discarded, excepting **A1** in which the $16d$ position refined to the unity.

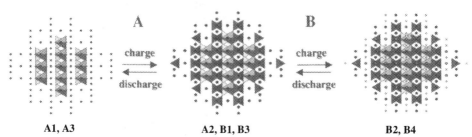

A1, A3   **A2, B1, B3**   **B2, B4**

Figure 5. View of the lithium-manganese spinel along the [221] direction with the Li tetrahedra aligned along [110]: grey and dark tetrahedra, at $8a$ and $8b$ respectively; o, at $16c$; and ●, at $16d$

In conclusion: i) the *8a* position is never fully occupied by Li, but always is partially shared with Mn; ii) during cycling Mn migrates from octahedral to tetrahedral and empty octahedral sites in a similar manner as it has been suggested [11] from ab initio calculations for the layered-to-spinel phase transition in Li_xMnO_2; iii) the structural changes associated to the Li deintercalation-intercalation during first cycle are practically reversible, only some Mn atoms remaining at *8a* positions; iv) the sole atom whose occupation tends to one when refining its population factor is that of oxygen at *32e*; v) three actual possibilities have been established; which of them is adopted closely depends on how far from the ideal spinel the starting material evolves; and vi) the migration of Mn finds a limit for each site of those it occupies. When this limit has to be surpassed, another position starts filling up. The effect, which remembers the successive filling of various levels in energy diagrams, can explain at atomic level the structural changes provoked by Li deintercalation-intercalation on $LiMn_2O_4$ cathodes in Li batteries.

Acknowledgements. This research was supported by the Spanish DGEIC, project Nos. PB97-0246, PB97-1200, MAT99-0892, MAT01-0562, and by the *Domingo Martínez* Foundation. The authors wish to thank Dr. Rosa M. Rojas and Mr. Jesús Merino for the DSC tests; and Mr. Enrique Gutiérrez-Monge for drawing figure 5.

REFERENCES

1. J. Akimoto, Y. Takahashi, Y. Gotoh, and S. Mizura, *Chem. Mater.* **12**, 3246 (2000).
2. H. Björk, T. Gustafsson, and J.O. Thomas, *Electrochem. Comm.* **3**, 187 (2001).
3. J. A. Campa, M. Velez, C. Cascales, E. Gutiérrez-Puebla, M.A. Monge, I. Rasines, and C. Ruíz-Valero, *J. Cryst. Growth*, **12**, 87 (1994).
4. A. Yamada, M. Tanaka, *Mater. Res. Bull.* **30**, 715 (1995).
5. M.A. Monge, E. Gutiérrez-Puebla, I. Rasines, and J.A. Campa in *Materials Science of Novel-Oxide-Based Electronics*, edited by D.S. Ginley, J.D. Perkins, H. Kawazoe, D.M. Newns, and A.B. Kozyrev (Mater. Res. Soc. Proc. **623**, Warrendale, PA, 2000), pp. 56-62.
6. T.Ohzuku, M. Kitagawa, and T. Hirai, *J. Electrochem. Soc.* **137**, 769 (1990).
7. Y. Gao, Y. and J.R. Dhan, *J. Electrochem. Soc.* **143**, 100 (1996).
8. M.R. Palacin, Y. Chabre, L. Dupont, M. Hervieu, P. Strobel, G. Rousse, C. Masquelier, M. Anne, G.G. Amatucci, and J.M. Tarascon, *J. Electrochem. Soc.* **147**, 845 (2000).
9. S. Geller, *Acta Crystall. B* **27**, 821 (1971).
10. J. Garrido, *Compt. Rend. Hebdom. Seanc. Acad. Sci. Paris* **200**, 69 (1935).
11. J. Reed, G. Ceder, and A. van der Ven, *Electrochem. Solid State Lett.* **4**, A78 (2001).

Mat. Res. Soc. Symp. Proc. Vol. 730 © 2002 Materials Research Society   V1.8

LiMBO₃ (M=Fe, Mn): Potential Cathode for Lithium Ion Batteries

Jan L. Allen, Kang Xu, Sam S. Zhang and T. Richard Jow
Sensors and Electron Devices Directorate, U.S. Army Research Laboratory
Adelphi, MD 20783-1197, U.S.A.

ABSTRACT

Recently discovered borates, LiMBO₃ (M=Fe, Mn), share similarities with LiFe(Mn)PO₄. They are polyanion structures, contain extractable lithium and suffer from low electronic conductivity. They are attractive to replace expensive, less abundant redox metals in current use in cathodes with environmentally friendly iron or manganese. Phosphate or borate groups adjacent to the redox active metal increase the voltage of the redox couple through an inductive effect. The LiFeBO₃ discharge curve shows a pseudo-plateau around 2.6 V for the Fe(II) / Fe(III) couple. This study brings to bear techniques to improve electrode conductivity to produce LiMBO₃ composite electrodes thus allowing access to some of the high, theoretical specific capacity. At low current, up to 70 percent of lithium could be extracted from LiFeBO₃ that was prepared in the presence of high surface area, highly electrically conductive carbon black. Attempts to improve the cathode properties of LiMnBO₃ were less successful.

INTRODUCTION

There is a growing interest in polyanion framework structured compounds as cathodes in lithium ion batteries. A polyanion containing a countercation with relatively high electronegativity functions as an electron-withdrawing group, withdrawing electron density from the covalent bonds of the nearest-neighbor redox cations. This inductive effect lowers the redox potential of the metal and raises the voltage of a cell with respect to the anode. Thus, in principle, one could tune the redox potential within a common structure by judicious choice of countercation. In practice, this inductive effect has been explored most intensively in the olivine, LiFePO₄ system [1]. The phosphate group raises the redox potential of the Fe(II) / Fe(III) couple to around 3.5 V. An iron-based cathode would have obvious benefits, as iron is cheap, abundant and environmentally benign.

There have been only a few reports exploring borate-containing compounds as lithium ion electrode materials [2-3]. Boron is lightweight, relatively abundant, and just slightly less electronegative than phosphorus. Recently, LiMBO₃ (M=Mn, Fe, Co) compounds have been reported [2]. Preliminary characterization of lithium extraction / insertion showed very little capacity (less than 0.04 lithium per formula unit) [2]. Recent work on LiFePO₄ has shown that careful synthesis to control particle size [4], the preparation of carbon-active nanocomposites [5] and post-synthesis carbon coatings [6] can dramatically improve cathode performance. We have implemented similar techniques to explore improvements in the performance of LiFeBO₃ and LiMnBO₃ cathodes. We have succeeded in improving initial capacity at low current to about 0.72 Li atoms per LiFeBO₃ formula unit, a specific capacity of 160 mAh/g but attempts to improve the cathodic behavior of LiMnBO₃ have been less successful. Greater improvements in cycling and performance at higher current will be needed for practical use of LiFeBO₃ as a cathode. The results emphasize the general importance of synthesis and electrode preparation methods.

EXPERIMENTAL DETAILS

Synthesis

$LiFeBO_3$ was first synthesized by the method of Leganeur et al. [2] by heating a stoichiometric mixture of $LiBO_2$ (Alfa) and $FeC_2O_4 \cdot 2H_2O$ (Alfa) at $350^{\circ}C$ for three hrs. followed by a $750^{\circ}C$ treatment for 15 hours under flowing N_2 in a tube furnace. Part of the product was thoroughly mixed with an acetone solution of cellulose acetate so that the $LiFeBO_3$/cellulose acetate pre-reaction ratio was 95:5. The acetone was evaporated under agitation. The cellulose acetate coated $LiFeBO_3$ was then heated under flowing N_2 at $400^{\circ}C$ for 3 hours to decompose the cellulose acetate to form a carbon-coated $LiFeBO_3$ sample and again at $750^{\circ}C$ in attempt to increase the conductivity of the carbon coating. The final carbon content was approximately 1wt.%. All subsequent syntheses used as starting materials Li_2CO_3 (Alfa), $FeC_2O_4 \cdot 2H_2O$ (Alfa) and H_3BO_3 (Aldrich) and all reactions were carried out under flowing N_2. A lower temperature route from these precursors was used to prepare $LiFeBO_3$ by using a temperature profile of $350^{\circ}C$ for 12hr and $550^{\circ}C$ for 24 hours. The synthesis in the presence of carbon was explored using as carbon source: steam activated carbon (Darco G60, Aldrich, typical surface area 900 m^2/g), a resorcinol-furaldehyde carbon aerogel precursor (RF aerogel, Marketech, Intl.), and highly electroconductive carbon black (Black Pearls 2000®, Cabot). The carbon additive was added to the starting materials so that the ratio of the $LiFeBO_3$ product to final carbon was 84:8. These carbon / reactant mixtures were ball-milled and then placed in a nickel combustion boat. The boats were placed in a tube furnace and the reactants were heated at $350^{\circ}C$ for 12 hours, then $800^{\circ}C$ for 24 hours in a flowing nitrogen atmosphere. The sample was furnace cooled to room temperature and then removed from the furnace. About 1% of the carbon was lost during the heating process.

$LiMnBO_3$ was first synthesized by the method of Leganeur [2] from a stoichiometric mixture of $LiBO_2$ (Alfa) and MnO(Alfa), though a higher reaction temperature and time was needed than reported by Leganeur. Perhaps this is a result of different source of starting reagents. The starting reagents were mixed by mortar and pestle and heated at $900^{\circ}C$ for 24 hours. Part of the sample was carbon coated using the cellulose acetate method as described for $LiFeBO_3$. $LiMnBO_3$ was also synthesized from a mixture of Li_2CO_3, MnC_2O_4 and H_3BO_3 in the presence of activated carbon as described for $LiFeBO_3$. Attempts to synthesize $LiMnBO_3$ in the presence of Black Pearls 2000® carbon black were not successful.

Characterization

Phase purity was evaluated by X-ray diffraction using a Philips Compact XRD system (PW1840) scanning from 10 to 90 degrees two-theta using an iron radiation source.

The composite electrodes were prepared by doctor-blading a PVDF (polyvinylidine fluoride, Polysciences Inc., Mol. Wgt. = 350,000) in NMP (1-methyl-2-pyrrolidinone, HPLC grade, Aldrich) slurry onto an aluminum foil current collector. The slurry was prepared by ball milling a 2% by weight solution of PVDF in NMP with the $LiMBO_3$ (M=Fe,Mn) or LiMBO3/C composite and additional carbon black (EC-600JD Ketjenblack®, Akzo Nobel), when needed, to produce a final cathode composition of 84% $LiFeBO_3$, 8% carbon and 8% PVDF binder. The

NMP was evaporated at room temperature in air and then the cathode was dried thoroughly under vacuum at 100°C.

Button cells were fabricated using Rayovac BR 2335 stainless steel hardware. One-half inch diameter cathodes were punched out from the coated aluminum foil and assembled with a one-half inch diameter lithium (Chemetall Foote Corporation, 0.63 mm thickness) anode and a ¼ inch diameter microporous polypropylene membrane as separator (Celgard® 2400). The electrolyte was a 1:1:3 by weight mixture of ethylene carbonate (Grant Chemical), propylene carbonate (Grant Chemical) and dimethyl carbonate (Grant Chemical) containing a 1 molal concentration of high-purity $LiPF_6$ (Stella Chemifa Corporation). Prior to use, the solvents were first dried over 3A molecular sieves (Aldrich) and then further dried over activated alumina (neutral, Brockman I, Aldrich). The solvent moisture level, determined by Karl-Fisher titration, was less than 12 ppm. Cells were assembled in either an argon-filled glove box (H_2O < 1ppm) or in a dry room with a dew point below −70°C. Cells were cycled between 4.2 and 2 V on a Maccor Series 4000.

RESULTS AND DISCUSSION

In a lithium ion battery, the cathode material is a donor of lithium ions. Lithium ions transfer charge through the electrolyte between the cathode and anode. In the first charging cycle, lithium is extracted from the cathode to compensate for the oxidation of a transition metal. For example, during the first charge of $LiFeBO_3$, lithium is extracted and iron is oxidized from iron(II) to iron(III). $LiFeBO_3$ and $LiMnBO_3$ both contain one lithium ion per formula unit. Thus, if one could extract all the lithium and completely oxidize iron (II) to iron(III) and manganese (II) to manganese(III) one can calculate a theoretical capacity of 220 and 222 mAh/g for $LiFeBO_3$ and $LiMnBO_3$, respectively. However, full capacity is rarely achieved in practical use owing to limitations of structure and the need for high ionic and electronic conductivity throughout the cathode composition as well as compromises made in order to achieve high cycle life.

Structural Description

The $LiFeBO_3$ structure is based upon chains of edge-sharing FeO_5 trigonal bipyramids. Each planar BO_3 group links three chains. Lithium occupies tetrahedral sites within the framework [2]. The h-$LiMnBO_3$ structure is based upon MnO_5 square pyramids, LiO_4 tetrahedron and BO_3 planar groups. The MnO_5 square pyramids share edges to form chains running along the c-axis. Each BO_3 group links three chains. Lithium occupies tetrahedral sites within the $[MnBO_3]_n$ framework [2]. Electronic conduction can occur through M-O-M bonding, unlike many polyanion structures, e.g., NASICON, in which the polyanions completely isolate each transition metal polyhedron.

LiFePO₄

Lithium iron phosphate was first described by Manitheram and Goodenough in 1991 as a good cathode material for low power applications as its performance was limited to low current and approximately 60% of the lithium could be extracted [1]. Armand et al. were the first to show that coating the LiFePO$_4$ with a thin, electrically conductive coating of carbon can dramatically improve the performance [6]. Nazar *et al.* have achieved almost full capacity at high rates through control of particle size and intimate carbon contact attained through mixing carbon or carbon precursor with the starting reagents [5]. Yamada has studied the optimization of the properties of LiFePO$_4$ through control of particle size by using lower reaction temperature during synthesis [4]. We have explored using some of these methods to improve the electrode performance of LiFeBO$_3$ and LiMnBO$_3$.

LiFeBO$_3$ and LiMnBO$_3$

LiMnBO$_3$ and LiFeBO$_3$ share common characteristics with LiFePO$_4$. They are polyanion-containing compounds with divalent transition metals and extractable lithium. The performance of LiMnBO$_3$ and LiFeBO$_3$ was evaluated in an electrochemical cell by Leganeur *et al.* Very little lithium was extracted (<0.04 per formula unit) [2]. We prepared these compounds by similar methods and obtained similar low capacity (ca. 0.03 per formula unit or 6 mAh/g). We have explored whether methods used to dramatically improve the performance of LiFePO$_4$ would also improve the performance of these borate compounds.

Carbon Coating

Armand *et al.* [6] have described a carbon coating method to produce electrode materials with increased surface conductivity. The method was applied to LiFePO$_4$ to dramatically improve its electrode performance. Based on this concept, we attempted to coat samples of LiFeBO$_3$ and LiMnBO$_3$ using a cellulose acetate precursor. The lithium extraction was increased to give a low current (0.05 mAh/ cm^2) capacity of about 20 mAh/g for both LiFeBO$_3$ and LiMnBO$_3$, too low for practical use.

Synthesis in Presence of Activated Carbon

Next, we explored the synthesis of the borate compounds in the presence of a high surface area carbon. We also used more reactive starting materials, Li$_2$CO$_3$ and H$_3$BO$_3$, in place of LiBO$_2$. First, for LiFeBO$_3$ we tried a series of reaction temperatures (from 500 to 900°C) and found that the relatively high temperature of 800°C gave the best results. We were able to extract 20% of lithium (40 mAh/g) at a low current of 0.05 mA/cm^2. Although, this is a five-fold increase in capacity it is still too low to be of any practical interest. Following a similar procedure for LiMnBO$_3$ gave a slight improvement in capacity to about 12 mAh/g.

Synthesis in Presence of a Carbon Precursor

Huang *et al.* [5] recently demonstrated that the preparation of nanocomposite cathodes through synthesis of LiFePO$_4$ in the presence of resorcinol-furaldehyde aerogel produces a cathode in which almost complete, high current rate extraction of lithium is achieved. Inspired by these results, we explored the preparation of a LiFeBO$_3$ / C nanocomposite through preparation of LiFeBO$_3$ in the presence of an RF aerogel. We were then able to achieve 30 mAh/g or Li$_{0.9}$FeBO$_3$, an improvement over conventional synthesis but not yet a practical capacity. The capacity of a LiMnBO$_3$ sample prepared by this method was about 20 mAh/g.

Synthesis in Presence of High Surface Area Carbon Black

Finally, we studied the preparation of $LiFeBO_3$ in the presence of a high surface area (1475 m^2/g), highly conductive carbon black, Black Pearls 2000®. The low current capacity was dramatically improved. We were able to achieve 160 mAh/g or $Li_{0.3}FeBO_3$. The first discharge curve is shown in figure 1. There is a semi-plateau at around 2.6 V. The capacity was obtained at 0.05 mAh / cm^2. Increased current rapidly reduces the capacity obtained. Much greater improvement will be needed in order to use the cathode in higher current applications. The fading behavior is shown in figure 2.

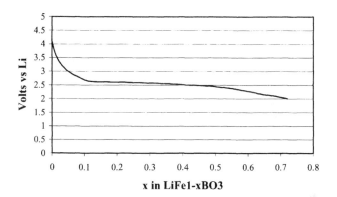

Figure 1. First discharge curve for a $LiFeBO_3$ sample prepared within an intimate mixture of high surface area carbon black (rate = 0.05 mAh/cm^2).

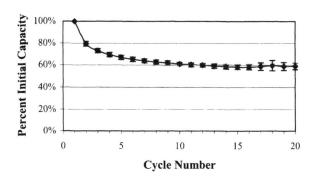

Figure 2. Retention of capacity over cycle number for $LiFeBO_3$ prepared within an intimate mixture of high surface area carbon black (rate = 0.05 mAh / cm^2).

There is a drop-off in capacity over the first 10 cycles after which the capacity stabilized at around 60% of the initial capacity (ca. 96 mAh/g). Unfortunately, we were unable to synthesize phase pure $LiMnBO_3$ in the presence of this carbon black.

CONCLUSIONS

$LiFeBO_3$ and $LiMnBO_3$ were studied as cathode materials in lithium batteries. The borate group raises the voltage of the Fe(II)/Fe(III) redox couple to around 2.6 V relative to Li. Methods to improve electrode performance were explored. By synthesizing $LiFeBO_3$ in the presence of high surface area carbon black, it was possible to electrochemically extract about 70% of the lithium, a specific capacity of 160 mAh/g. Attempts to improve the capacity of $LiMnBO_3$ were less successful.

REFERENCES

1. A. Manitheram and J.B. Goodenough, J. Solid State Chem. 71, 349 (1991).
2. V. Leganeur, Y. An, A. Mosbh, R. Portal, A. Le Gal La Salle, A. Verbaere, D. Guyomard and Y. Piffard, Solid State Ionics, 139, 37 (2001).
3. J.L.C. Rowsell, J. Gaubicher and L.F. Nazar, J. Power Sources 97-98, 254 (2001).
4. A Yamada, S.C. Chung and K. Hinokuma, J. Electrochem. Soc. 148(3) A224 (2001).
5. H. Huang, S.-C. Yin and L.F. Nazar, Electrochem. Solid-State Letters, 4(10) A170 (2001).
6. Michel Armand, Simon Besner, Nathalie Ravet, Martin Simoneau, Alain Vallee and Jean-Francois Magnan, European Patent No. EP 1 049 182 (2 November 2000).

Mat. Res. Soc. Symp. Proc. Vol. 730 © 2002 Materials Research Society

Energy Storage Materials Based on Iron Phosphate

Pier Paolo Prosini, Maria Carewska, Marida Lisi, Stefano Passerini, Silvera Scaccia, and Mauro Pasquali[1]
ENEA, C.R. Casaccia. Via Anguillarese 301, 00060 Roma, Italy.
[1]Dipartimento ICMMPM, Facoltà di Ingegneria, Università di Roma "La Sapienza", Roma, Italy.

ABSTRACT

Several iron phosphates were synthesized by solution-based techniques and tested as cathodes in non aqueous lithium cells. The addition of phosphate ions to a solution of iron (II) produced crystalline $Fe_3(PO_4)_2$. This material is easily oxidized by air to form an amorphous phase that is able to reversibly intercalate lithium. The amorphous compound was identified to be a mixture of $FePO_4$ and Fe_2O_3. A new synthetic route was developed to prepare pure amorphous $FePO_4$. Amorphous $LiFePO_4$ was obtained by chemical lithiation of $FePO_4$. The material was heated at 500°C under reducing atmosphere to obtain nano-crystalline $LiFePO_4$. This latter material showed excellent electrochemical performance when used as cathode of lithium cells.

INTRODUCTION

Iron is a very attractive material to be used in the field of lithium battery, especially to build large size batteries for powering electric vehicles or for realizing dispersed electrical power sources. However, for a series of reasons, iron and its derivatives have not met with success as cathode materials. In fact, in the iron-based oxides containing O^{2-} as anion, the Fe^{4+}/Fe^{3+} redox energy tends to lie too far below the Fermi energy with respect to a lithium anode, while the Fe^{3+}/Fe^{2+} couple is too close to it.

The use of polyanions such as XO_n^{m-} (X= Mo, W, S, P, As) has been shown to lower the Fe^{3+}/Fe^{2+} redox energy to useful levels. The effect of the structure on the Fe^{3+}/Fe^{2+} redox energy of several iron phosphates was investigated by Padhi et al. [1]. Among these, lithium iron phosphate (ordered olivine-type structure) is emerging as the most promising material to replace lithiated transition metal oxides as cathode material for secondary lithium-ion batteries [2-4].

In this paper we report the synthesis of several iron phosphates and their behavior as cathodes in lithium batteries.

EXPERIMENTAL DETAILS

Crystalline $Fe_3(PO_4)_2$ was prepared by spontaneous precipitation from iron (II) and phosphate aqueous solutions. A 0.06 M $Fe(NH_4)_2(SO_4)_2 \cdot 6H_2O$ (Carlo Erba, RPE) solution was added at ambient temperature to a constantly stirred 0.04 solution of K_2HPO_4 (Carlo Erba, RPE), in a 1:1 volume proportion. A pale-blue gel started to form after the addition was completed. The gel was collected on a membrane filter, washed and dried. Amorphous iron phosphate was obtained by oxidation of crystalline $Fe_3(PO_4)_2$ in air at 100°C. After the heating treatment, the color of the powder changed from white to dark-yellow.

Amorphous $FePO_4$ was synthesized by spontaneous precipitation from iron (II) phosphate aqueous solutions, using hydrogen peroxide as oxidizing agent. In practice, to a solution of 0.025 M $Fe(NH_4)_2(SO_4)_2 \cdot 6H_2O$ was added an equimolar solution of $NH_4H_2PO_4$, in a 1:1 volume proportion. Then 3mL of concentrated hydrogen peroxide solution was added to the solution at ambient temperature under vigorous stirring. A white precipitate started to form immediately after the addition of hydrogen peroxide.

Amorphous $LiFePO_4$ was obtained by chemical lithiation of amorphous $FePO_4$ by using LiI as reducing agent. Amorphous $FePO_4$ was suspended in a 1M solution of LiI in acetonitrile. The suspension was kept under stirring for 24 h, filtered on a membrane filter (0.8 μm), washed several times with acetonitrile and dried under vacuum.

Crystalline $LiFePO_4$ was obtained by heating the latter (amorphous) compound in a tubular furnace at 550°C for 1h under reducing atmosphere ($Ar/H_2=95/5$).

Composite cathode tapes were made by roll milling a mixture of the iron phosphate with the binder (Teflon, DuPont) and the carbon black (SuperP, MMM Carbon) which was used to increase the electronic conductivity. Electrodes were punched in form of discs, with a typical diameter of 8 mm. The electrodes were assembled in a sealed cell formed by a polypropylene T-type pipe connector with three cylindrical stainless steel (SS316) current collectors. Lithium foils were used both as anode and reference electrode, and a glass fiber was used as separator. The cells were filled with ethylene carbonate /diethyl carbonate 1:1 $LiPF_6$ 1M electrolyte solution. The cycling tests were carried out automatically by means of a battery cycler (Maccor 2000). Composite cathode preparation, cell assembly, test and storage were performed in a dry room (R.H. < 0.2% at 20°C).

DISCUSSION

Nazar et al. [4] showed that the reduction of the grain size could be one of the possible routes to enhance the performance of $LiFePO_4$ to make feasible its use as cathode in high-power density lithium-ion batteries.

To prepare $LiFePO_4$ with smaller grain size we added phosphate ions to a solution of iron (II) sulfate in presence of lithium ions. As a result we obtained the formation of a pale-blue gel. The chemical analysis of the precipitate showed the presence of iron and phosphate in a 3:2 ratio but not the presence of alkali or sulfate ions. The XRD analyses confirmed that the compound was crystalline $Fe_3(PO_4)_2$. The compound was found very sensible to oxidation. After heating in air at 100°C, the compound transformed to form an amorphous phase. Thermo gravimetric analysis, Mössbauer spectroscopy, and XRD spectra suggested that the oxidized compound is a mixture of $FePO_4$ and Fe_2O_3 in a 4:1 molar ratio [5]. Figure 1 is a SEM micrograph of the oxidized material. It is characterized by thin layers, arranged in alternating and overlapping planes, spreading out radially as a flower corolla. The material was tested as a cathode in a lithium cell. Figure 2 shows the voltage profiles for different discharge currents. The rechargeability of the material is illustrated in Fig. 3, where the specific capacity of the cathode (based on the weight of active material) is reported *vs.* the cycle number.

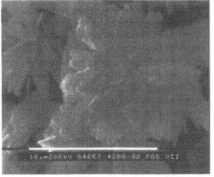

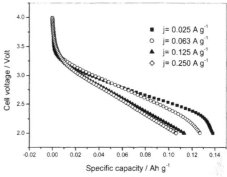

Figure 1. SEM micrograph of oxidized $Fe_3(PO_4)_2$.

Figure 2. Voltage profiles for oxidized $Fe_3(PO_4)_2$ as a function of the cycle number.

At the lowest current used the cell was able to deliver a specific capacity of 0.138 Ah g^{-1}. By increasing the current, the utilization of the active material decreased, and about 0.105 Ah g^{-1} were delivered by the cell discharged at a current density that was 30 times higher. The lithium intercalation into the amorphous structure was seen to be very reversible. The capacity fade was about 0.1 % per cycle. The excellent performance of the amorphous iron(III)phosphate was related to the unique microstrutture and the small particle size achieved by the solution based synthesis. Since iron(III)oxide that is present as a by-product is electrochemically non active for

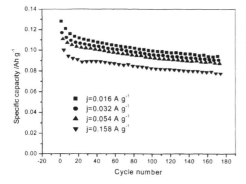

Figure 3. Specific capacity for oxidized $Fe_3(PO_4)_2$ as a function of the cycle number.

Figure 4. SEM micrograph of amorphous $FePO_4$.

lithium intercalation in the voltage range investigated (4.0-2.0 V), the electrochemical properties of the material can be completely ascribed to the amorphous iron(III)phosphate phase.

For this reason we investigated a new synthetic route to prepare pure amorphous iron(III)phosphate. The new route involved the spontaneous precipitation from equimolar aqueous solutions of $Fe(NH_4)_2(SO_4)_2 \cdot 6H_2O$ and $NH_4H_2PO_4$, using hydrogen peroxide as oxidizing agent [6]. Figure 4 shows a SEM micrograph of the material obtained after precipitation. The material is characterized by a sponge-like structure where is difficult to recognize any structural organization. A sample heated at 400°C was chosen to evaluate the capacity as a function of the discharge rate and the capacity retention as a function of prolonged cyclation. The cell was cycled galvanostatically under various discharge rate with a cut-off voltage of 2.0 V. The cell was always recharged with the same charge procedure, to assure identical initial conditions; a first constant current step at 0.17 A g^{-1} (C rate) was applied until the voltage reached 4.0V followed by a constant voltage step until the current lowered below 0.017 A g^{-1} (C/10 rate). Figure 5 shows the voltage profiles recorded at various discharge rates. At the lowest current density used (0.017 A g^{-1}), the material was able to deliver a specific capacity of 0.108 Ah g^{-1}. By increasing the current density the active material utilization decreased and about 0.08 Ah g^{-1} were delivered when discharging the cell at C rate. The rechargeability of the material at different discharge rates is shown in Fig. 6. The capacity fade, evaluated during cycling at the lowest discharge rate, was about 0.075% per cycle.

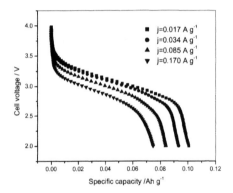 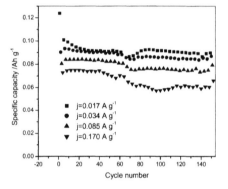

Figure 5. Voltage profiles for amorphous FePO$_4$ at different discharge currents.

Figure 6. Specific capacity for amorphous FePO$_4$ as a function of the cycle number.

Amorphous LiFePO$_4$ was obtained by chemical lithiation of amorphous FePO$_4$ by using LiI as reducing agent and the correspondent crystalline phase was obtained by heating amorphous LiFePO$_4$ at 550°C for 1h under reducing atmosphere. Figure 7 and 8 show SEM micrographs of the lithiated materials before and after the heating treatment. The materials are characterized by a globular structure with a grain size of about 100-150 nm. Figure 9 shows the voltage profiles as a function of the specific capacity at different discharge rates. The cell was discharged galvanostatically under different specific currents, ranged from 0.017 up to 0.510 A g^{-1}. The discharge and charge cut-off voltages were 2.0V and 4.0V, respectively. The cell was always recharged at the same specific current (0.017 A g^{-1}), to assure identical initial conditions for any discharge. At the lowest specific discharge current used (0.017 A g^{-1}), the cell was able to deliver a specific capacity 0.162 Ah g^{-1} based on the active material weight only, in a discharge time of about 10 h. By increasing the current density the utilization of the active material decreased, and about 0.140 Ah g^{-1} were delivered in about 0.29 h at a specific current 30 times

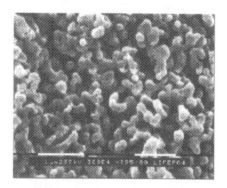

Figure 7. SEM micrograph of amorphous LiFePO$_4$.

Figure 8. SEM micrograph of crystalline LiFePO$_4$.

larger. This result appears very good when compared with lithium iron phosphate synthesized by traditional solid-state chemistry, in which an increase in the discharge current results in a severe capacity fading [2]. The cycle life of the material is illustrated in Fig. 10, where the cathode specific capacity (based on the weight of the active material) is reported *vs.* the cycle number. The insertion/release process was driven at 0.170 A g^{-1} between fixed voltage values (2.0/4.0 V *vs.* Li). The specific capacity slowly decreased upon cycling. The capacity fade was evaluated about 0.25% per cycle.

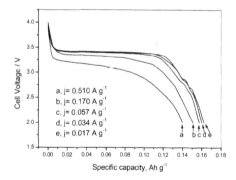

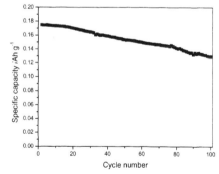

Figure 9. Voltage profiles for nano-crystal line LiFePO$_4$ at different discharge currents.

Figure 10. Specific capacity as a function of the cycle number.

CONCLUSIONS

Several iron phosphates were produced by solution-based methods. This latter approach allows to obtain materials characterized by smaller grain size. The materials were tested as cathodes in lithium cells. They showed large specific capacity and low capacity fading. The excellent performance of the compounds was related to the unique microstructure and the small particle size achieved by the solution based synthesis. The inherent structure of the materials is retained throughout the entire intercalation range and the fatigue and the stress that arise from the lithium insertion/release process can be well accommodated into the structure. The wet chemical method provide an effective-cost preparation method and, in addition to the fact that iron and phosphorus are inexpensive and not toxic materials, could lead to a considerable reduction of the battery price.

ACKNOWLDGMENTS

This work is part of the activities sponsored by the Italian Ministry of University and Scientific Research.

REFERENCES

1. A.K. Padhi, K.S. Nanjundaswamy, C. Masquelier, S. Okada, and J.B. Goodenough, *J. Electrochem. Soc.*, **144**, 1609 (1997).
2. A.K. Padhi, K.S. Nanjundaswamy, and J.B. Goodenough, *J. Electrochem. Soc.*, **144**, 1188 (1997).
3. A. Yamada, S.C. Chung, and K. Hinokuma, *J. Electrochem. Soc.,* **148**, A224 (2001).
4. H. Huang, S.-C. Yin, and L.F. Nazar, *Electrochem. Solid State Lett.*, **4**, A170 (2001).
5. P.P. Prosini, L. Cianchi, G. Spina, M. Lisi, S. Scaccia, M. Carewska, C. Minarini, and M. Pasquali, *J. Electrochem. Soc.*, **148**, A1125 (2001).
6. S. Scaccia, M. Carewska, A. Di Bartolomeo, and P.P. Prosini, *Thermochim. Acta*, **383**, 145 (2002).

Mat. Res. Soc. Symp. Proc. Vol. 730 © 2002 Materials Research Society   V1.11

Preparation and Characterization of Tin/Carbon Composites for Lithium-Ion Cells

Ronald A. Guidotti[1], David J. Irvin[2], William R. Even, Jr.[3], and Karl Gross[3]
[1]Sandia National Laboratories, P.O. Box 5800, Albuquerque, NM 87185-0614
[2]Naval Air Warfare Center, Polymer Science and Engineering Branch, China Lake, CA 93555
[3]Sandia National Laboratories, P.O. Box 969, Livermore, CA 94551

ABSTRACT

A number of Sn/C composites were prepared for evaluation as anode materials for Li-ion cells. In one case, samples were prepared by incorporation of Sn species into organic precursors that were then pyrolyzed under an Ar/H_2 cover gas to prepare the Sn/C composites. They were also prepared by decoration of various types of carbon with nanoparticles of Sn by electroless deposition using hydrazine. The carbons examined included a disordered carbon prepared in house from poly(methacrylonitrile), a mesocarbon microbead (MCMB) carbon, and a platelet graphite. The Sn/C composites were examined by x-ray diffraction (XRD) and scanning electron microscopy (SEM) and were also analyzed for Sn content. They were then tested as anodes in three-electrode cells against Li metal using 1M $LiPF_6$ in ethylene carbonate (EC)/dimethyl carbonate (DMC) solution. The best overall electrochemical performance was obtained with a Sn/C composite made by electroless deposition of 10% Sn onto platelet graphite.

INTRODUCTION

There has been increasing interest in recent years in the development of improved materials for use as anodes in lithium-ion cells. Both synthetic and natural graphites as well as disordered carbons have been studied for this purpose. Graphite has a maximum theoretical capacity of 372 mAh/g of carbon, but the disordered carbons can have capacities two or more times that. Unfortunately, the disordered carbons also exhibit high irreversible, first-cycle loss. A number of transition metals and transition-metal alloys have theoretical capacities much greater than that of graphite. Sn, for example, can alloy with Li to form $Li_{4.4}Sn$, which has a theoretical capacity of 994 mAh/g of Sn. However, such materials show rapid fade with cycling due to attrition of the alloy caused by large volume changes. One way of mitigating this problem is to use the metal on a support or substrate. With combination of Sn or Sn alloys with graphite, it is possible to realize the capacities of both materials when used as lithium-ion anodes [1,2].

In this work, we report on the preparation and characterization of Sn/C composites made with platelet, spheroidal, and disordered carbon. Elemental Sn was deposited on the surfaces of the carbons by chemical reduction in both nonaqueous and aqueous solutions. In some cases, the metal salts were incorporated into the carbon precursors prior to pyrolysis. The Sn/C composites were tested over a range of current densities as lithium-ion anodes.

EXPERIMENTAL

Preparation of Sn/C Composites

Sn/C Composites by Pyrolysis – Tin dibutyl bis(2,4-pentane-dionate) [$SnBu_2(acac)_2$] (Gelest, Inc.) was dissolved in dioxane and shaken to form a particulate suspension with thermally stabilized powdered precursor of poly(methacrylonitrile) (PMAN)/divinyl benzene (DVB) polymer (PMAN/DVB = 3:1 m/m) for 3 days. (The cellular-carbon precursor was prepared by an inverse emulsion polymerization [3].) The slurries were precipitated by adding drop-wise into liquid nitrogen and the resulting solids were freeze-dried over 3 days. This procedure yielded ~30 g of very finely dispersed tin samples that were pyrolyzed under heating schedule #1 of Table I. This yielded ~7g of each Sn/C composite sample with target values of 10% and 15% Sn.

The following tin compounds, tin (II) oxalate (SnOx), tin (II) acetate (SnAc$_2$), and dimethyl tin oxide (Me$_2$SnO) (all Gelest, Inc.) were dissolved in a solution of HAc/2 M HCl (3/1 v/v) and were shaken with thermally stabilized PMAN/DVB polymer for 3 days. The slurries were precipitated into liquid nitrogen and the resulting solids were freeze-dried over 4 days. This procedure yielded ~30 g of each sample to be pyrolyzed under heating schedule #2 of Table I. This yielded ~7g of each Sn/C composite sample with target values of 10% and 15% Sn.

Table I. Heating schedules use for preparation of Sn/C composites.

Heating Schedule #1	Heating Schedule #2
25°C to 300°C @ 2°C/min	25°C to 300°C @ 2°C/min
Hold 2 h	Hold 2 h
300°C to 350°C @ 1°C/min	300°C to 350°C @ 1°C/min
Hold 6.7 h	Hold 5 h
300°C to 760°C @ 2°C/min	300°C to 765°C 2°C/min
Hold 5 h	Hold 5 h
Cool to 300°C @ 3°C/min	Cool to ambient
Cool to ambient	--------

Sn/C Composites by Electroless Deposition – Tin bis-(2,4-pentane dionate) [tin acetylacetonate or Sn(acac)$_2$] (Gelest, Inc.) was dissolved in dimethyl sulfoxide and shaken with PMAN carbon (pyrolyzed at 800°C), KS-6 graphite (Lonza), and MCMB-2528 carbon (Osaka Gas Corp.) for three days. Two equivalents of anhydrous hydrazine were added to the slurries and the mixtures were shaken for 25 h. The slurries were then precipitated into diethyl ether. After collection of solids by filtration, they were washed with diethyl ether and then dried under vacuum for 24 h to yield ~10 g of Sn/C composites with a target value of 10% Sn. These materials were not thermally treated further.

Sn(acac)$_2$ reduces to tin metal and 2,4-pentanedione (eqn. 1). The solvent and byproducts are removed under vacuum to yield the Sn/C composite without pyrolysis.

$$\text{(Sn(acac)}_2\text{)} \xrightarrow{N_2H_4} Sn + N_2 + 2 \text{ (pentanedione)} + 2H^+ \tag{1}$$

Electrochemical Testing

A ½" polyfluoroalkoxy (PFA) Swagelok® tee cell was used for electrochemical characterization of the samples. The samples were "doctor bladed" onto a copper foil and contained 85% carbon, 15% polyvinylidenedifluoride (PVDF) as a binder and 5% Super 'S' carbon black as a conductive additive. The electrolyte was 1M $LiPF_6$ in EC/DMC (1:1 v/v) (Merck). The water level was typically <40 ppm as determined by Karl Fischer titration.

The samples were subjected to the test protocol shown in Table II. The first 20 cycles gives a measure of the tendency for fading. The remaining cycles provide information on the rate capability. The same rates were used for intercalation and deintercalation

Table II. Testing protocol for 32-cycle galvanostatic cycling between 2 V and 0.01 V.

Number of Cycles	Intercalation Rate	Deintercalation Rate
20	C/5	C/5
3	C/10	C/10
3	C/2.5	C/2.5
3	C/1.25	C/1.25
3	C/0.625	C/0.625

RESULTS AND DISCUSSION

Physical and Chemical Properties – The compositions of the Sn/C composites are summarized in Table III. Because of the relatively low pyrolysis temperatures, a considerable quantity of N, O, and H remained in the PMAN carbons from the pyrolysis tests. These were higher than the values found for the 800°C PMAN used for the electroless-plating experiments. The loadings of Sn on the electroless-plated carbons of 8.4% to 10.1% were close to the target value of 10% Sn. However, all of the Sn levels in the Sn/C synthesized by pyrolysis of PMAN/DVB precursors were much lower than their target values. In the case of SnOx, the Sn loadings were <0.3%. This suggests volatilization of the SnOx during the heating and pyrolysis. Similar results were noted for Me_2SnO. Somewhat higher Sn loadings (~1%) were realized using $SnAc_2$. The highest Sn levels were obtained with the $Sn(acac)_2$. Levels of only 5.6% and 6.5% were obtained for target values of 10% and 15% Sn, respectively.

The XRD pattern of the pyrolyzed samples showed narrow Sn diffraction lines from the presence of micro-spheres (crystallites >100 nm) of metallic tin; there was no evidence for tin oxides. By contrast, materials prepared by electroless deposition of Sn were devoid of strong Sn Bragg reflections, indicative of either amorphous Sn or nanostructured (<100 nm) Sn crystallites. The scanning electron microscopy (SEM) photomicrograph of the Sn/C composite made by pyrolysis using the $SnAc_2$ precursor is shown in Figure 1a. Corresponding information for the Sn/C composite for the Me_2SnO is presented in Figure 1b. Large spheres of Sn were evident and show that the Sn had melted, since the pyrolysis temperatures were well above the melting point of Sn (231.9°C). Thus, any nanoparticles of Sn present initially consolidated upon melting to form these large spheres. Similar results were observed for the precursors made with $SnBu_2(acac)_2$. No elemental Sn was found in the composite when the $SnAc_2$ precursor was used. SEM photomicrographs of the Sn/C composites made by electroless plating are presented in Figures 2a, 2b, and 2c. At this magnification, it was not possible to readily observe the discrete Sn nanoparticles.

Table III. Elemental analysis of Sn/C composites.

Precursor	Target % Sn	Found % Sn	% C[*]	% H[*]	% N[*]	% O[*]
$SnBu_2(acac)_2$	10	5.58	82.54	1.78	3.57	12.12
$SnBu_2(acac)_2$	15	6.47	81.61	1.41	3.37	13.61
SnOx	10	0.13	83.11	1.37	3.65	11.87
SnOx	15	0.27	79.56	1.78	3.34	15.31
$SnAc_2$	10	1.04	83.55	1.43	3.59	11.43
$SnAc_2$	15	0.98	82.27	2.16	3.75	11.83
Me_2SnO	10	0.45	82.890	1.23	3.44	12.54
800°C PMAN	10	8.42	85.99	0.76	1.68	11.56
KS-6	10	10.14	94.28	<0.6	<0.6	~4
MCMB-2528	10	8.71	94.81	<0.6	<0.6	~4

* Composition of carbon portion of Sn/C composites.

Electrochemical Properties – The performance of pyrolyzed Sn/C PMAN composites based on the $SnAc_2$ and the $SnBu_2(acac)_2$ precursors are compared to the Sn-coated PMAN carbon in Figure 3a. The PMAN derived from $SnAc_2$ serves as a pseudo-control, since it contained very little Sn. However the incorporation of 5.6% Sn into the PMAN matrix in the case of the composite made using $SnBu_2(acac)$ did not impact the capacity for the first three cycles. Afterward, however, the fade (loss in capacity with cycling) was higher and the rate capability dropped rapidly. When the PMAN was coated with a somewhat higher loading of Sn (8.42%), the negative impact was even more severe and the capacity and rate capability dropped by over 40% by the 20th cycle. The presence of Sn impeded the lithiation/delithiation process in some undefined manner.

The performance of the Sn-coated MCMB-2528 carbon is shown in Figure 3b. Similar data for the KS-6 graphite is shown in Figure 4a. In contrast to the PMAN and MCMB carbons, the platelet graphite benefited greatly by the Sn coating. The combined theoretical capacities of the

1μm 20000X   2μm 10000X

Figure 1. SEM photomicrographs of Sn/C composites for precursors of a) tin acetate (1 μm marker) and b) dimethyl tin oxide (2 μm marker).

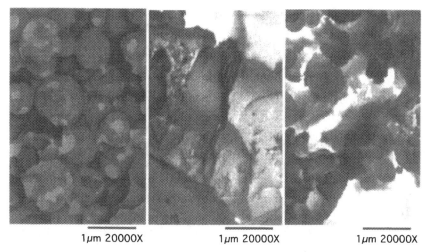

Figure 2. SEM photomicrographs of Sn/C composites made by electroless deposition on a) PMAN carbon, b) KS-6 graphite, and c) MCMB-2528 carbon (1 μm markers).

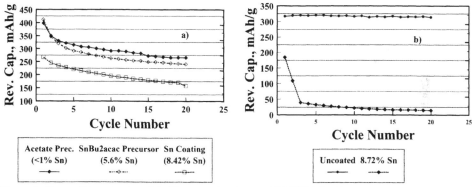

Figure 3. Reversible capacity of Sn/C composites in 1M LiPF6/EC-DMC at C/5 rate from 2V to 0.01 V for a) pyrolyzed PMAN and coated 800°C PMAN and b) coated and uncoated MCMB-2528.

Sn ($Li_{22}Sn_4$) and the graphite (LiC_6) were realized for a total composite capacity of 450 mAh/g. The overall improvement in the rate capability was not as great, however.

The relative performance of the coated carbons is summarized in Figure 4b. These data show how the morphology and structure of the carbon can have a dramatic impact on the relative performance when used as anodes in Li-ion cells. The graphite coated with 10% Sn cycled with little capacity loss after 20 cycles, while the 800°C carbon coated with 8.42% Sn showed a much lower capacity with a gradual fade. The worst performer was the MCMB-2528 coated with 8.72% Sn. It lost most of its capacity after only three cycles. It also had poor rate capability.

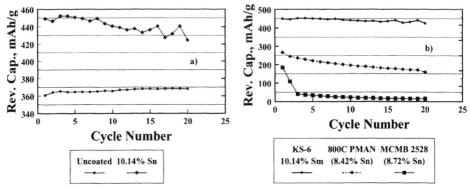

Figure 4. a) Reversible capacity of Sn-coated KS-6 in 1M LiPF6/EC-DMC at C/5 rate from 2V to 0.01 V and b) relative comparison of performance among Sn-coated carbons.

CONCLUSIONS

Sn/C composites containing micron-sized spheres of Sn were prepared by ramp-heating to 760° - 765°C under Ar/5% H_2 of PMAN/DVB precursors containing a various Sn salts [(SnOx, Me_2SnO, $SnAc_2$, and $SnBu_2(acac)_2$]. Levels of Sn of 5.6% and 6.5% are obtained when targeting 10% and 15% Sn using $SnBu_2(acac)_2$ but <1% with the other Sn precursors. These materials do not cycle as well as Li-ion anodes as the 800°C control PMAN carbon. Electroless deposition of 8.4% Sn nanoparticles onto disordered PMAN carbon using $Sn(acac)_2$ reduces the reversible capacity and rate capability by interfering with intercalation in some undefined way. Even worse results are obtained with spherical MCMB-2528 carbon at this level. By electroless deposition of 10% Sn nanoparticles onto KS-6 platelet graphite, the full capacity of the graphite and Sn of 450 mAh/g total composite is realized. This composite shows good cycling at a C/5 rate and at higher rates as well. More work in this area is merited.

REFERENCES

1. B. Veeraraghavan, A. Durairajan, R.E. White, B. N. Popov, and Ronald Guidotti, "Synthesis and Characterization of Sn-Coated SFG10 Graphites as Negative Electrodes in Li-Ion Cells," *198th Meeting of The Electrochemical Society*, Phoenix, AZ, October 22-27, 2000.
2. D. J. Irvin, W. R. Even, K. Gross, and R. A. Guidotti, "Tin/Carbon Composites: Crystalline Sphere to Amorphous Powders," SAND2001-8397 (August 2001).
3. W. R. Even and D. P. Gregory, *MRS Bull.*, **19**(4), 29 (1994).

ACKNOWLEDGMENTS

The pyrolysis experiments were conducted by Marion Hunter; Herb Case performed the electrochemical testing. Funding for this program was provided in part by Lockheed Martin through its Shared Vision program. Sandia National Laboratories is a multiprogram laboratory operated by Sandia Corp., a Lockheed Martin company, for the United States Department of Energy under Contract DE-AC04-94AL85000.

Mat. Res. Soc. Symp. Proc. Vol. 730 © 2002 Materials Research Society V1.12

Synthesis and Processing of High Capacity, High Cycle life and High Discharge Rate Defective Manganospinel films for Rechargeable batteries

Deepika Singh*, Heinrich Hofmann*, Won-Seok Kim†, Valentin Craciun†, Rajiv K. Singh†

*Powder Technology Laboratory (LTP), Department of Materials Science, Swiss Federal Institute of Technology (EPFL), CH 1015 Lausanne, Switzerland.

†Department of Materials Science and Engineering, University of Florida, Gainesville, Florida 32611, USA

ABSTRACT

Significant effort to develop a robust rechargeable battery has been put in the past two decades. The efforts were mainly focused on developing rechargeable battery systems which exhibit high capacity, long cycle life and high discharge rate capabilities. $LiMn_2O_4$ based cathodes have been researched extensively as they are not only economical but also environmentally desirable. Research includes composition and doping variation, formation of novel phases and microstructural tailoring, but none of the material modifications have successfully satisfied all the above mentioned performance criteria. In this paper we show a correlation between processing parameters, microstructure and electrochemical performance of Li-Mn-O cathode films. In addition we discuss the formation of metastable oxygen-rich lithium manganospinels, using a unique ultraviolet assisted deposition process. These defective films exhibit high capacity (> 230 mAh/gm), long cycle life (less than 0.05 % capacity loss per cycle for the first 700 cycles), and high discharge rates (> 25 C for 25 % capacity loss). The long cycle life and high capacity was attributed to the ability to cycle the Mn^+ valence to less than 3.5 without onset of Jahn-Teller structural transformation, while the high discharge rate was attributed to the extremely high diffusivity of Li^+ in the defective $Li_{1-\delta}Mn_{2-2\delta}O_4$ phase.

INTRODUCTION

Efforts have been focused on replacing the conventional $LiCoO_2$ positive electrodes with cheaper, safer and environmentally friendly materials such as $LiMn_2O_4$ and related compounds[1-7]. In $LiMn_2O_4$ phase, the extraction of a Li^+ ion from the tetrahedral sites takes place in two closely spaced steps at approximately 3.9 ~ 4.2 V vs. Li / Li^+ ($LiMn_2O_4 \rightarrow Mn_2O_4$ (λ-MnO_2)), whereas the insertion of a Li^+ ion into the octahedral sites occurs at approximately 3 V vs Li / Li^+ ($LiMn_2O_4 \rightarrow Li_2Mn_2O_4$). The insertion of lithium into $LiMn_2O_4$ is naturally accomplished by a reduction of the average oxidation state of manganese from 3.5 to 3. The presence of more than 50 % of Jahn-Teller ions (Mn^{3+}) in the host structure introduces a cubic to tetragonal distortion (from c/a = 1 to c/a = 1.16), which upon repeated cycles is believed to deteriorate the electrical contact and decrease the capacity of the cathode[8, 9]. Thus, the maximum usable capacity of $LiMn_2O_4$ is limited to 0.5 Li atom per Mn atom, which translates to the maximum useable capacities of 120 ~ 140 mAh/gm. The cycle life (defined by 75% reduction in capacity) is typically in the range of 200 ~ 400 cycles, whereas the maximum discharge rate is limited by the diffusivity of lithium into the positive cathode. Intense efforts in the past decade to simultaneously enhance the capacity, discharge rate and cycle life in the past decade have met with limited success[4-7]. For example, whereas high capacities (exceeding 200 mAh/gm) have been observed in nanocrystalline Li-Mn-O and LiMnO$_2$ materials, these materials have shown very low discharge rates or short cyclelife[4, 6]. On the other hand, high discharge rate nanostructured cathode materials have been developed, however, the total capacities in these systems are typically not adequate[7].

In this paper we show for the first time that the formation of defective Li$_{1-\delta}$Mn$_{2-2\delta}$O$_4$ manganospinels, which contain vacancies at both tetrahedral lithium sites and octahedral manganese sites can lead to high capacity (>230 mAh/gm), high cycle life (>700 cycles) and high discharge-rates (>25 'C'). Such compounds also are characterized by a Li/Mn ratio of 0.5 and have an average Mn$^+$ valence state varying from 3.5 to 4.0 (depending on the value of δ). For a value of δ = 0.11 this compound has a stoichiometric form of Li$_2$Mn$_4$O$_9$ with a Mn$^+$ oxidation state of 4.0[10]. The higher the value of δ, the lower the capacity at 4 V, the smaller the lattice parameter, and the better cyclability in the 3 V region. Although it has been speculated that the oxygen-rich lithium manganospinels such as Li$_2$Mn$_4$O$_9$ can deliver high steady capacities in excess of 150 mAh/gm, the reproducible synthesis of fully oxidized single phase using bulk solid state chemistry technique is quite difficult[10-12]. Strict control of the experimental conditions such as temperature, time, particle size and oxygen partial pressure have not led to production of fully oxidized phase material[10, 13, 14]. Increased oxygen incorporation has particularly been difficulty as higher processing temperature reverts the defective spinel back to the stoichiometric LiMn$_2$O$_4$ phase[5]. Thin film deposition techniques, which have typically been used, have not been successful in maintaining a constant stoichiometric Li/Mn ratio or enhancing the oxygen content above that in the bulk counterparts[6].

EXPERIMENTAL

We used pulsed laser deposition (PLD) and the ultraviolet-assisted pulsed laser deposition (UVPLD) process to synthesize Li-Mn-O films having excellent electrochemical characteristics. In the UVPLD process an ultraviolet lamp emits significant radiation at 185 nm which helps to break the molecular oxygen in the deposition chamber into atomic oxygen and other reactive species such as ozone[14, 15]. The enhanced reactivity of non-equilibrium oxygen species leads to the formation of Li$_{1-\delta}$Mn$_{2-2\delta}$O$_4$ films during the UVPLD process. It is also well known that the pulsed laser deposition process helps to maintain the stoichiometry of the films primarily because of the rapid ablation process and the relatively high partial pressure of oxygen in the chamber[16]. Many processing parameters such as substrate temperatures and oxygen pressure were varied to obtain different microstructures. In previous publications we demonstrated that the use of an ultraviolet assisted deposition process leads to enhanced oxygen incorporation in several oxide based systems including Y$_2$O$_3$, ZrO$_2$, BaSrTiO$_3$, LaCaMnO$_3$, and related systems[14, 15, 17].

RESULTS AND DISCUSSION

Figure 1 compares the X-ray diffraction (XRD) spectra of films deposited onto silicon by the pulsed laser deposition (PLD) versus the UVPLD technique at the same processing temperature (600 °C) and oxygen pressure (1 mbar). The figure shows that the x-ray diffraction peaks are qualitatively quite similar for both spectra, except that the peaks in the UVPLD film much sharper indicating a higher degree of crystallinity.

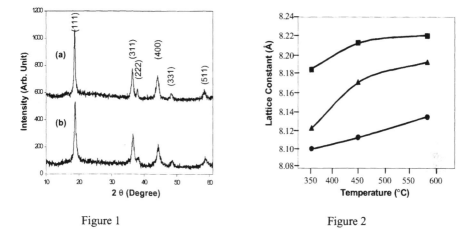

Figure 1   Figure 2

Figure 1: XRD patterns of LiMn$_2$O$_4$ films deposited at 600 °C and an oxygen pressure of 1 mb using a) PLD. b)UVPLD.

Figure 2: Lattice parameter of the thin films as a function of temperature. Films were deposited at an O$_2$ pressure of 1 mbar by PLD on silicon (squares), by UVPLD on silicon (triangles) and by the UVPLD on stainless steel (circles). The lattice parameters were calculated from the peaks positions of the (400) and (111) Bragg peaks.

A more significant difference obtained from the X-ray diffraction patterns was the variation in the lattice parameter as a function of processing temperature, as shown in Figure 2. The figure shows that the PLD films have a lattice parameter in the range of 8.18 ~ 8.22 Å which corresponds to the LiMn$_2$O$_4$ phase. However, the films deposited on silicon and stainless steel by the UVPLD process under the same temperatures exhibit a much smaller lattice parameter when compared to PLD films. As the Li/Mn ratio measured by Nuclear Reaction Analysis and Rutherford Backscattering Spectroscopy was close to 0.5 for all films, the smaller lattice parameter indicates the formation of the oxygen-rich Li$_{1-\delta}$Mn$_{2-2\delta}$O$_4$ spinel[18]. Further confirmation of the Li$_{1-\delta}$Mn$_{2-2\delta}$O$_4$ phase was obtained from XPS studies which showed that the atomic concentration of Mn^{4+}/Mn^{3+} and O/Mn were in the range of > 3.0, and 2.2 - 2.3, respectively for UVPLD films. It should also be noted that the lattice parameter of UVPLD films on steel substrate is smaller than films deposited on silicon substrate because of the higher compressive stress generated in the films due to thermal expansion mismatch between the film and the substrate. If one takes into account the thermal expansion effect (thermal expansion coefficient of Si = 4 × 10^{-6} /K and stainless steel = 15 × 10^{-6} /K), the lattice parameters of UVPLD films on silicon and stainless steel approximately match each other. Studies in the literatures have suggested that oxygen-rich spinels are stable at temperature below 400 °C[5, 11].

However, we envisage that the presence of atomic oxygen species in our experiments may increase the stability temperature for $Li_{1-\delta}Mn_{2-2\delta}O_4$ phase.

Extensive electrochemical and battery measurements were conducted on the $LiMn_2O_4$ and $Li_{1-\delta}Mn_{2-2\delta}O_4$ films synthesized by PLD and UVPLD techniques, respectively. The electrochemical measurements were conducted in a coin cell configuration using 1M $LiPF_6$ in EC-DMC electrolyte. The cyclic voltammogram(CV) of Li/ $Li_{1-\delta}Mn_{2-2\delta}O_4$ cell cycled from 2.2 V to 4.6 V is shown in the Figure 3.

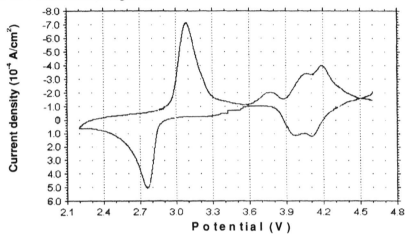

Figure 3: CV of Li $_{1-\delta}$ Mn $_{2-2\delta}$O$_4$ film. The film was deposited by UVPLD at 625 C and an O$_2$ Pressure of 1 mb. Scan rate was 0.1mv/s.

The CV spectrum shows that the lithiation and delithiation reactions are reversible. For the defective spinel, during anodic scan lithium ions are inserted at approximately 3.1 V whereas the remaining lithium ions are inserted in 2 step processes at 4.05 and 4.19 V, respectively. The redox peaks were used to estimate the lithium ion diffusivity using the Randle-Sevick equation. Diffusivity values of $5.0 \times 10^{-7} \sim 2 \times 10^{-8}$ cm^2/sec were obtained from $Li_{1-\delta}Mn_{2-2\delta}O_4$ films, which is $1 \sim 2$ order of magnitude higher than the diffusivity values obtained from $LiMn_2O_4$ materials[5]. It should also be noted that unlike $LiMn_2O_4$ films, the 3 V capacity is much larger than the 4 V capacity which is characteristic of the $Li_{1-\delta}Mn_{2-2\delta}O_4$ oxygen rich spinels.

The capacity, cycle life and the maximum discharge rate capability were determined for $Li_{1-\delta}Mn_{2-2\delta}O_4$ films which were approximately 2.0 μm in thickness. Figure 4 show the cycle life of the $Li_{1-\delta}Mn_{2-2\delta}O_4$ films deposited on steel substrate at 400 °C and 1 mbar for films cycled in both 4 V (4.5 ~ 3.5 V) and 4 and 3 V (4.5 ~ 2.5 V) ranges. It should be noted that films deposited at higher temperature exhibited much higher capacities but a slightly reduced cycle life. For comparison the cycling characteristics of $LiMn_2O_4$ films are also shown. These films were cycled at 1000 μA/cm^2, which corresponds, to approximately a 10 C rate. The initial capacities of the $Li_{1-\delta}Mn_{2-2\delta}O_4$ films is approximately 80 mAh/gm and 230 mAh/gm when cycled in the 4.5 - 3.5 V and 4.5 - 2.5 V ranges, respectively. Under extended cycling conditions in both these voltage ranges, excellent cycle life was obtained. In the 4 V range, less than 15 % capacity loss is obtained when cycled for over 1300 cycles whereas in both 3 V and 4 V range, the capacity loss is approximately 30 % when cycled to more than 700 cycles. In contrast, typical $LiMn_2O_4$ films exhibit very short cycle life as expected when subjected to 3 V cycling conditions.

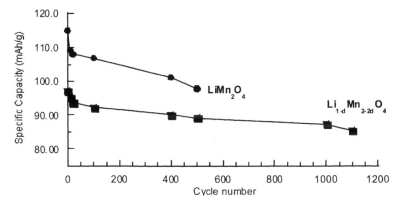

Figure 4: Cycle life of Li-Mn-O films. The charge and discharge rate of 1 mA/cm^2, which is equivalent to 10 C rate, was used.

The high capacity and the excellent cycle life of $Li_{1-\delta}Mn_{2-2\delta}O_4$ thin film cathodes may be attributed to a number of factors. Relatively low cycle life in bulk $LiMn_2O_4$ electrodes has been attributed to the dissolution of Mn from the cathode, inhomogeneous local structure and Jahn-Teller transition which occurs when the average valence state of Mn in $LiMn_2O_4$ is 3.5[5, 9]. Our results suggest that during 4 V cycles, the average valence of Mn in the films is less than 3.5, however we still do not observe significant degradation in the electrochemical characteristics. We attribute the long cycle life due to specific thin film effects namely: (i) presence of compressive stresses, (ii) high film homogeneity and (iii) formation of oxygen rich $Li_{1-\delta}Mn_{2-2\delta}O_4$ phase. The films deposited on steel substrates have compressive strains of approximately 0.6 % to 1 % as indicated by the reduced lattice parameter. The compressive stresses may prevent the onset of the Jahn-Teller transition in these films. The films are very homogenous with strong grain boundary contact and lack of binder and conducting phases. These effects combined with a relatively highly defective structure and a higher oxidation state of manganese of $Li_{1-\delta}Mn_{2-2\delta}O_4$ may prevent the onset of Jahn-Teller structural transition and better accommodate stress during cycling.

Another important characteristic of the battery is the effect of the discharge rate on the battery capacity. Our results indicate that if the microstructure and the film thickness are carefully tailored, very high rate discharge capability are obtained. The films were discharged at high rates in both the 4 V and 3 V regions. Very high discharge rate capabilities are obtained from $Li_{1-\delta}Mn_{2-2\delta}O_4$ for both in the 4 V and 3 V cycling. For example at a discharge rate of 25 C, the capacity degradation is less than 25 % in the 4 V and 3 V regions. Even when discharge rates of 50 C in the 3 V region, nearly 60 % of the capacity is still available for use. In contrast, the $LiMn_2O_4$ spinels show much higher capacity losses when discharged at high 'C' rates. We attribute the high rate discharge capability of $Li_{1-\delta}Mn_{2-2\delta}O_4$ to rapid intercalation kinetics of the lithium ions in the $Li_{1-\delta}Mn_{2-2\delta}O_4$ films. The large number of vacancies in the 8a tetrahedral and 16d octahedral sites, combined with a large number of line defects such as grain boundaries may significantly enhance the lithium ion diffusion coefficient. In conclusion, we have demonstrated that the synthesis of oxygen-rich $Li_{1-\delta}Mn_{2-2\delta}O_4$ thin films by ultraviolet assisted method provide excellent, unmatched battery characteristics for potential applications in consumer electronics, energy storage, and high discharge rate applications.

ACKNOWLEDGEMENT

Part of this work was supported by the US DOE and the Swiss Federal Office of Energy, Switzerland.

REFERENCES

1. Tarascon, J. M. & Armand, M, *Nature* 414, 359-367 (2001).
2. Amatucci, G. G., Pereira, N., Zheng, T. & Tarascon, J. M., *J. Electrochem. Soc.* 148, A171-A182 (2001).
3. Thackeray, M. M., David, W. I. F., Bruce, P. G. & Goodenough, J. B., *Mater. Res. Bull.* 18, 461-472 (1983).
4. Chiang, Y. M., Sadoway, D. R., Jang, Y. I., Huang, B. & Wang, H., *Electrochem. Solid-State Lett.* 2, 107-110 (1999).
5. Xia, Y. & Yoshio, M., *J. Electrochem. Soc.* 144, 4186-4194 (1997).
6. Bates, J. B. *et al.* Prefered orientation of polycrystalline $LiCoO_2$ films. *J. Electrochem. Soc.* 147, 59-70 (2000).
7. Che, G., Lakshmi, B., Fisher, E. & Martin, C, *Nature* 393, 346-349 (1998).
8. Guyomard, D. & Tarascon, J. M, *J. Electrochem. Soc.* 139, 937-947 (1992).
9. Thackeray, M. M. , *J. Electrochem. Soc.* 142, 2558-2563 (1995).
10. Masquelier, C. *et al.* , *J. Solid. State Chem.* 123, 225-266 (1996).
11. Kilroy, W. P., Ferrando, W. A. & Dallek, S. , *J. Power Sources* 97-98, 336-343 (2001).
12. Kock, A. de et al. Defect spinel in the system $Li_2O \cdot yMnO_2(y>2.5)$: , *Mater. Res. Bull.* 25, 657-664 (1990).
13. Gummow, R. J., Kock, A. de. & Thackeray, M. M., *Solid State Ionics* 69, 59-67 (1994).
14. Craciun, V. & Singh, R. K, *Electrochem. Solid-State Lett.* 2, 446-447 (1999).
15. Craciun, V. & Singh, R. K., *Appl. Phys. Lett.* 76, 1932-1934 (2000).
16. Singh, R. K. & Narayan, J. , *Phys. Rev. B* 41, 8843-8859 (1990).
17. Kumar, D. & Singh, R. K. *J. Vac. Sci. & Techn. A* (in press) (2002).
18. Singh, R. K. *et al. J. Electrochem. Soc.* (submitted).

Correspondence and requests for materials should be addressed to D. Singh (e-mail: Singh@sinmat.com)

Hydrogen Fuel Cells and
Hydrogen Storage

Mat. Res. Soc. Symp. Proc. Vol. 730 © 2002 Materials Research Society   V2.2

Understanding the Formation Energy of Transition Metal Hydrides

Huw J. Smithson[1,2], Dane Morgan[1], Anton Van der Ven[1], Chris Marianetti[1], Ashley Pedith[1] and Gerbrand Ceder[1]
[1]Department of Materials Science and Engineering, Massachusetts Institute of Technology, Cambridge, MA 02139, U.S.A.
[2]Department of Materials Science and Metallurgy, University of Cambridge, Cambridge, CB2 3QZ, U.K

ABSTRACT

A detailed analysis of the formation energies of transition metal hydrides is presented. The hydriding energies are computed for various crystal structures using Density Functional Theory. The process of hydride formation is broken down into three consecutive, hypothetical reactions in order to analyse the different energy contributions, and explain the observed trends. We find that the stability of the host metal is very significant in determining the formation energy, thereby providing a more fundamental justification for Miedema's "law of inverse stability" [1] (the more stable the metal, the less stable the hydride). The conversion of the host metal to the structure formed by the metal ions in the hydride (fcc in most cases) is only significant for metals with a strong bcc preference such as V and Cr - this lowers the driving force for hydride formation. The final contribution is the chemical bonding between the hydrogen and the metal. This is the only contribution that is negative, and hence favourable to hydride formation. We find that it is dominated by the position of the Fermi level in the host metal.

INTRODUCTION

The absorption of hydrogen in materials is an important phenomenon. In many metals it can lead to hydrogen embrittlement - causing premature failure under stress [2]. There is a renewed interest in metal hydrides as a potential hydrogen source for small portable fuel cells. Reversible hydriding can be used as a fuel storage mechanism for operation in automobiles or for stand-alone power generation.

Metals for hydrogen storage need to be able to form hydrides with a high hydrogen-to-metal ratio, but not be too stable, so that the hydrogen can be easily released without excessive heating. Magnesium and magnesium-nickel hydrides contain a relatively high fraction of hydrogen by weight, but need to be heated to at least 250 or 300°C in order to release the hydrogen.

Understanding the stability of metal hydrides is essential in order to rationally investigate and design potential hydrogen-storage materials. Our research aims to explain the stability of hydrides on the basis of their electronic structure. Using modern first principles energy computations, the energy for hydride formation is calculated for a large number of metals. By systematically breaking down the hydride formation energy into several components, the trend across the transition metals is explained. It is demonstrated that the electronic structure of the host metal is a key factor in determining the stability of the hydride.

COMPUTATIONAL DETAILS

All computations were performed in the Local Density Approximation (LDA) to Density Functional Theory. Atomic cores are represented with ultra-soft pseudopotentials as implemented in the Vienna Ab Initio Simulation Package (VASP) [3,4]. An energy cut-off of 300eV was used for all hydride calculations and k-point sampling was performed on a 12x12x12 grid for all structures. All structures were fully relaxed (cell parameters as well as internal geometries) so as to obtain the minimum of the total energy. All the calculations performed here are non-spin-polarised, even for cases known to be magnetic. We have performed spin-polarised calculations (LSDA) for a few hydrides and found no qualitative changes.

RESULTS AND DISCUSSION

Formation energy of *3d* transition metal hydrides

This report focuses on the *3d* transition metals. For each metal we looked at a range of hydride structures, including those known to be stable for these metals. A plot of the formation energies of the *3d* transition metals is shown in Figure 1.

Looking at Figure 1 it is clear that all of the structures share a similar trend across the period. In order to interpret this trend the formation energy is broken down into three hypothetical, consecutive reactions, the energy of each of which can be directly related to the electronic structure. The sum of these three reaction energies will be the hydride formation energy. Since most of the structures show the same trend across the transition metal series, we focused on hydrides with the fluorite (CaF_2) and rock salt (NaCl) structures for this analysis. Both of these structures can be described in terms of an fcc metal lattice with hydrogen atoms occupying either all the octahedral interstices (NaCl) or all the tetrahedral interstices (CaF_2).

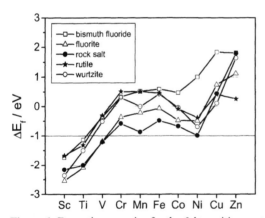

Figure 1. Formation energies for the *3d* transition metal hydrides in various structures. (Energy given per H_2 molecule).

The three reactions are:
1. Conversion of the metal from its equilibrium structure to fcc (which is the arrangement of the metal ions in the fluorite and rock salt structures). This is referred to as the *structural* effect as it is determined by the topology of the metal ions in the hydride structure.
2. Expansion of the metal fcc structure to the lattice parameter of the hydride. This is referred to as the *elastic* effect.
3. Introducing the hydrogen atoms into the interstices of the fcc lattice so as to form the hydride. This is the *chemical* effect and is a measure of the bonding between the hydrogen and the metal.

The structural effect is only important if the stable metal structure is bcc (because fcc and hcp are close in energy). So this is only a significant factor in metals with a strong bcc preference such as V and Cr. The structural energy changes for the *3d* transition metals are shown in Figure 2.

A large (but positive) contribution to the hydride formation energy comes from step 2 - the expansion of the metal lattice to the lattice parameter of the hydride. This elastic energy contribution, shown in Figure 3, increases as one moves through the *3d* series up to about Fe and then decreases slightly. An increase in the energy cost for stretching the metal leads to a reduction in the stability of the hydride. For the fluorite structure, the variation in this contribution to the hydride formation energy is nearly one and a half electron volts, indicating that it may contribute a substantial part to the variation in hydride energies observed in Figure 1.

The energy of the third step - the insertion of hydrogen - is shown in Figure 4. This energy contribution is strongly negative and is the energetic reason for hydride formation. The energy gain resulting from hydride insertion decreases when moving through the transition metal series - in agreement with much of the trend observed in Figure 1.

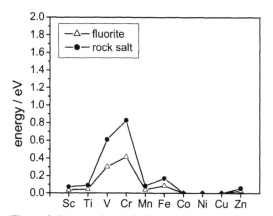

Figure 2. Structural contribution to formation energy for *3d* transition metal hydrides in fluorite and rock salt structures. (Energy given per H_2 molecule).

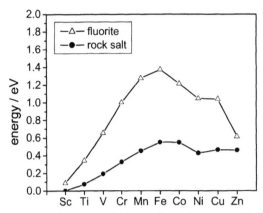

Figure 3. Elastic contribution to formation energy for *3d* transition metal hydrides in fluorite and rock salt structures. (Energy given per H_2 molecule).

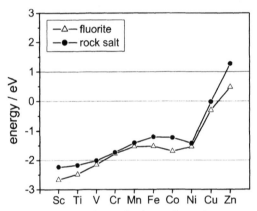

Figure 4. Chemical contribution to formation energy for *3d* transition metal hydrides in fluorite and rock salt structures. (Energy given per H_2 molecule).

Interpretation of results

The three components of energy can all be understood in terms of the band structure and charge transfer within the material. However, a full treatment of this is significantly more complex than the scope of this brief paper allows. The main points are presented below.

The effects of the structural energy contribution are small compared to the others, and the effect is only significant for V and Cr - the two bcc metals. For this reason we shall concentrate on the elastic and chemical contributions.

Given how much emphasis has been put in the past on the chemical / electronic interaction of hydrogen with the transition metal, it is surprising that such a large contribution to the trend in formation energy actually arises from the expansion of the metal lattice to adopt the hydride lattice parameter. The elastic penalty for hydrogen insertion is largely determined by the cohesive energy of the metal. For transition metals the cohesive energy is dominated by the d-electrons. For early transition metals, the Fermi level lies in strongly bonding levels near the bottom of the d-band. However, there are few d-electrons in these early transition metals so that the cohesive energy is rather low. This translates to a high formation energy for the hydride, as there is only a small elastic penalty to pay for hydrogen insertion. Going to the right in the $3d$ series, more bonding levels get filled, increasing the cohesive energy, hence reducing the magnitude of the hydride formation energy. The late transition metals have the Fermi level in the anti-bonding states near the top of the d-band and the cohesive energy decreases with increasing electron count. For the late transition metals such as Cu, the s-states also have to be considered to understand the cohesive energy.

The chemical interaction energy, shown in Figure 4, varies significantly across the transition metals. It is clearly responsible for the lack of stability of Cu and Zn hydrides. The strong bonding between hydrogen and the metal is due to the overlap of hydrogen-s and metal-s orbitals, creating a deep bonding band. This energy contribution is similar for most transition metals. The variation of this energy component across the metal series is likely to be determined by the electron added to the d-states of the metal due to the s-to-d promotion of metal electrons (an in-depth analysis showed that on adding hydrogen to a transition metal, the hydrogen $1s$ states replaced the metal $4s$ states, accommodating one of the metal $4s$ electrons, and promoting the other into the d-band). We believe that there are two main competing factors which contribute to changes in s-to-d promotion energy as a function of atomic number. Because the d-orbitals fill with increasing atomic number, the electron must be promoted to a higher energy orbital. However, this effect is counteracted by an overall lowering of the d-orbital energies associated with the increasing nuclear charge. These terms are relatively closely balanced, with the destabilising effect of d-band filling winning out from Sc to Fe and the effect of the increasing nuclear charge slightly dominating from Fe to Ni. The dramatic increase in hydride formation energy after Ni is due to the radical rearrangement of the band structure associated with the filled d-bands of Cu and Zn. The promoted electron is forced into a much higher energy orbital. This is very clear in Zn, where the hydride has 13 valence electrons and only 12 states available from the metal-d and metal-hydrogen bonding orbitals. The extra electron is promoted to a very high energy band, destabilising the hydride.

The trends in the three components to hydride formation can now be put together to explain the formation energies in Figure 1. The early transition metals suffer little loss of cohesive energy from the metal-metal bonds as hydrogen is inserted, but benefit from large direct hydrogen-metal interactions. This gives a large negative hydriding energy for these metals. Moving to the systems with more d-electrons, the energy cost of deforming the metal lattice increases and the chemical interaction between hydrogen and metal decreases, so that the stability of the hydride decreases. The apparent local minimum in hydride energy at Mn is due to the strong increase in energy of hydride formation for V and Cr, which are stable bcc metals. For these systems, forming either rock salt or fluorite hydrides comes with the strong cost of rearranging the metal lattice into the less favourable fcc structure. For the hydrides of the later transition metals Co and Ni the formation energy decreases again and the hydrides become somewhat more stable. This is due to a reduction in the contributions from the structural and elastic energies, and a fairly flat chemical contribution. However, the metal cohesive energy (and therefore the structural energy) remains quite large, making for an overall weak driving force for hydride formation. For Cu and Zn the hydriding energy again rises sharply, driven by the sharp increase in chemical energy.

Our observation that the cohesive energy of the metal plays a key role in the formation energy of the hydride gives a more fundamental justification for Miedema's "law of inverse stability" [1]. Miedema observed that the more stable the metal, the less stable the hydride.

CONCLUSIONS

With these results it is now possible to set about designing a hydride with appropriate stability for hydrogen storage. Both the cohesive energy of a metal and the position of the Fermi level - key factors for hydride stability - can be easily manipulated by alloying. The chemical effect can be influenced somewhat by tailoring the position and density of states at the Fermi level through doping with metals that have either high or low electron to atom ratio. A more effective way of influencing the formation energy of hydrides may be through the cohesive energy of the host metal. As demonstrated, this is a large factor in the stability of the hydride. For designing hydrides with marginal stability (so that hydrogen can be released easily) this may be the most promising direction, as the cohesive energy of alloys can be varied over wide ranges through compositional modification. In addition, this is a quantity that can be rapidly predicted for a large number of alloys with the first principles methods used in this work.

Obviously, it should be realised that the formation energy is only one important parameter for a hydride to be useful as a hydrogen source or for reversible hydrogen storage. Hydrogen diffusion, maximal hydrogen content, oxidation resistance, cost and weight are also important factors that would need to be considered in designing an ideal material for hydrogen storage.

ACKNOWLEDGEMENTS

This work was funded through the Laboratory for Energy and the Environment at the Massachusetts Institute of Technology and by the Department of Energy under contract number DE-FG0296ER45571. Computing resources were provided by the National Partnership for Advanced Computing Infrastructure (NPACI). HJS acknowledges support from the Cambridge-MIT Institute for supporting his stay at MIT. This research was supported in part by NSF cooperative agreement ACI-9619020 through computing resources provided by the National partnership for Advanced Computational Infrastructure at the San Diego Supercomputer Center.

1. A. R. Miedema, *J. Less Common Metals* **32**, 117 (1973).
2. H. Vehoff, in *Hydrogen in Metals III, Properties and Applications*, Edited by H. Wipf (Springer Verlag, Berling-Heidelberg, 1997), vol. 73, p. 215.
3. G. Kresse, J. Furthmüller, *Comput. Mat. Sci.* **6**, 15-50 (1996).
4. G. Kresse, J. Hafner, *Phys. Rev. B.* **47**, 558 (1993).

Mat. Res. Soc. Symp. Proc. Vol. 730 © 2002 Materials Research Society

Hydrogen Absorption over Li – Carbon Complexes

J.Z. Luo, P. Chen, Z.T. Xiong, K.L. Tan, and J.Y. Lin
Department of Physics, National University of Singapore, 10 Kent Ridge Crescent, Singapore,
119260

ABSTRACT

A remarkable reduction in reaction temperature was found for the hydrogenation of Li metal in
Li-C mixture. H_2 uptake started at 50°C, became vigorous at 150°C and slowed down at
temperatures above 200°C. *In-situ* XRD characterizations revealed that Li-C intercalation
compounds such as LiC_6 and LiC_{12} existed in the Li-C samples, and LiH formed after the
hydrogenation taking place. Increasing the carbon content in the Li-C mixture, from Li/C = 10:1
to 5:1 and then to 2:1, would enhance the reactivity of hydrogenation accordingly. Carbon
nanotubes, with smaller size and larger specific area, showed even greater enhancement of the
hydrogenation of lithium metal than graphite. The mechanism for the low temperature
hydrogenation of Li-C samples was studied and discussed.

INTRODUCTION

Hydrogen is considered as an alternative energy source for the 21 century. The utilization of
hydrogen energy, particular for vehicles, requires the on-board hydrogen storage capacity to be
competitive with today's petroleum [1]. The development of hydrogen storage materials with
high efficiency has attracted much interest and effort worldwide [2-4]. In order to meet the
practical requirements, the material should be able to perform hydrogenation/dehydrogenation
quickly under moderate temperatures and pressures. The existing hydrogen storage systems,
especially those with relatively high hydrogen storage capacity, i.e. Li and Mg hydrides or
physico-sorption by activated carbon etc., either operate at high temperatures or at cryogenic
conditions (liquid nitrogen temperature). Recently, we studied the absorption of hydrogen over
mixtures of M-graphite (M = Li, Na, or K) and found that the hydrogenation temperature for Li
and K were dramatically reduced by the presence of carbon. In this paper, we extend the research
to study the effect of temperature, carbon content, and carbon source on the hydrogenation of the
Li-C system. In addition, a more detailed work on the understanding of the hydrogenation
mechanism was performed. Our results revealed that Li-C intercalation compounds act as active
centres for H_2 adsorption and dissociation, and are responsible for the high hydrogen reactivity
of Li-C samples.

EXPERIMENTAL

The Li-C samples were prepared by mechanical mixing. Lithium metal (99%, granulated with
2.5 mm diameter, Fluka) and graphite (99.9%, ~ 4 μm, Merck) were hammered in a mortar for
10 mins. Then the mixture was pressed into pellets under 5,000 kgcm^{-2} pressure and cut into
small pieces (2-3 mm) for further use. To avoid the contamination of O_2, N_2, and moisture,
sample preparation was done in Ar (99.9 %) atmosphere. Carbon nanotubes were prepared and
purified according to ref. [5].

Pressure Composition Isotherm (PCI) unit was commercially provided by Advanced Materials Co. (USA). Two methods were applied to study the hydrogenation behaviours of Li-C samples. One was autosoak method, in which 420 psig of purified H_2 was introduced into sample chamber and temperature was increased from room temperature to 350°C at a heating rate of 1°C min⁻¹. The other was autopci method, in which the temperature was kept at desired point and the H_2 pressure was increased from 0 to 550 psig step by step. 500 mg of sample was used each time. The structure of samples was determined by a X-ray diffreactrometer (XRD, Bruker D8-advanced) with Cu ka ($\lambda = 0.15406$ nm) radiation source.

RESULTS

Figure 1 shows the autosoak profiles of lithium metal, Li-graphite (10:1) and Li-CNTs (10:1), respectively. The unit of Y-axis is the molar ratio of H/Li which reflects the hydrogenation degree of sample. The maximum value of H/Li = 1 should be achieved if all of lithium metal in the sample converted to LiH:

$$2Li + H_2 \rightarrow LiH \qquad\qquad \Delta H = -175.6 \text{ KJ mol}^{-1}$$

For lithium metal (see Fig. 1a), no hydrogenation took place at temperatures lower than 500°C. As temperatures approaching 650°C, the reaction accelerated vigorously. At 720°C, most of lithium converted to LiH (H/Li = 0.95). In contrast to metal lithium, Li-graphite and Li-CNTs can adsorb H_2 at much lower temperatures but with relatively slow kinetics. For Li-graphite (10:1) (see Fig. 1b), hydrogen adsorption set out at 50°C then accelerated continuously with the raising of temperature. From 150 °C to 200°C, H/Li increased from 0.1 to 0.35. The hydrogenation slowed down at temperatures above 200°C. At 350°C, the molar ratio of H/Li is 0.43. Similar tendency can be observed over Li-CNTs (10:1) sample except the higher H/Li troughout every temperature point. At 350°C, the molar ratio of H/Li is 0.54.

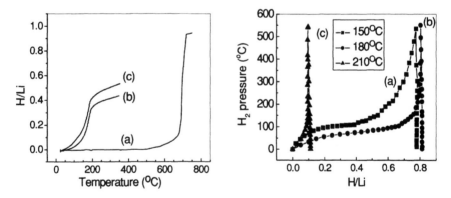

Figure 1 (left): Autosoak profiles of (a) metal Li, (b) Li-graphite (10:1), and (c) Li-CNTs (10:1). The hydrogen pressure was 420 psig. The heating rate is 1°C/min.
Figure 2 (right): PCI profiles of Li-graphite (10:1) at various temperatures.

To understand the temperature effect on the hydrogenation of Li-C samples, PCI measurements were performed at 150°C, 180°C, and 210°C, respectively. The results are shown in Fig. 2. At 150°C and 180°C, the PCI profiles resemble to each other and also the PCI curves of typical metal-hydride systems; a plateau pressure exists. However, their hydrogenation behaviours are somewhat different. At 150°C (curve a in Fig. 2), the average plateau pressure is 110 psig. The H/Li is 0.78 when pressure increases to 550 psig. At 180°C, the plateau pressure is 69 psig, a little lower than that at 150°C. H/Li is 0.81 at 550 psig, also a slightly higher than that at 150°C. Interestingly, the hydrogenation of Li-graphite sample seems to be depressed at 210°C. It can be found that the H/Li remains below 0.12 even when H_2 pressure increase to 550 psig.

The carbon content also affects the hydrogenation of the Li-C samples. The PCI measurements conducted on Li-graphite sample with Li/C = 10:1, 5:1, and 2:1, respectively, showed that the plateau pressure decreased from 69 to 19 and to 12 psig, and the maximum H/Li increased from 0.81 to 0.90 and to 0.96, respectively (figure 3). All of these results were summarized in Table 1. Compared with graphite, carbon nanotubes show an even stronger enhancement in the hydrogenation of lithium metal. As shown in Figure 3, the plateau pressure for Li-CNT (10:1) is only 30 psig, while for Li-graphite with same Li/C ratio, it is 69 psig.

Table 1. PCI results for various Li-C samples.

Sample	H_2 Pressure at plateau (Psig)*	Maximum X (X= H/Li)	Formula of the complex	H capacity (wt%)
Li-graphite (10:1) (150°C)	110	0.78	$Li_{10}C_1H_{7.8}$	9.6
Li-graphite (10:1) (180°C)	69	0.81	$Li_{10}C_1H_{8.1}$	9.9
Li-graphite (10:1) (210°C)	-	0.09	$Li_{10}C_1H_{0.9}$	1.1
Li-graphite (5:1) (180°C)	19	0.90	$Li_5C_1H_{4.5}$	7.0
Li-graphite (2:1) (180°C)	12	0.96	$Li_2C_1H_{1.92}$	4.9
Li-CNTs (10:1) (180°C)	30	0.79	$Li_{10}C_1H_{7.9}$	9.7

(*) Usually, the H_2 pressure were obtained at H/Li = 0.4 in the PCI profiles.

The mechanism of hydrogenation over Li-C sample was studied by *in-situ* XRD. As the maximum pressure for *in-situ* XRD cell is 14.7 psig, we chose Li-graphite sample with Li/C = 2/1 for XRD measurement. As shown in figure 4, there are LiC_6 (2θ = 23.9°, 48.8°), weak LiC_{12} (2θ = 25.5°, 51.7°), and lithium metal (2θ = 35.9°) phases in the freshly made sample [6]. After introducing hydrogen at 180°C, new phase attributed to LiH (2θ = 37.95°, 44.1°, and 63.7°) appeared (figure 4a and 4b) revealing the occurrence of hydrogenation under these experimental conditions. Prolonging the reaction time, the peak intensity of LiH phase increased gradually. Concurrently, those of LiC_6 and Li metal decreased continuously. However, the phase of lithium metal weakened quickly. As can be seen from figure 4, the peak intensity of lithium metal reduced by 75% after 1 hour of reaction, while it is only 1/10 for LiC_6 (figure 4a and 4b). The complete hydrogenation of lithium within LiC_n (n=6 or12) structure need longer time. After 24 hours reaction, almost all of lithium converted to LiH, graphite (positioned at 2θ = 26.2°, 53.9°) regenerated (figure 4c and 4d).

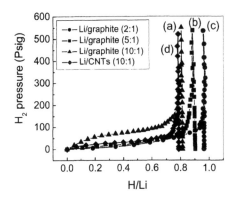 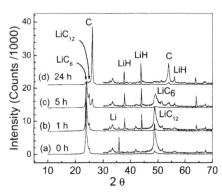

Figure 3 (left): PCI profiles of Li-graphite with Li:C molar ratio of (a) 10:1, (b) 5:1, (c) 2:1, and (d) Li-CNTs (10:1) at 180°C.

Figure 4 (right): *In-situ* XRD patterns of Li-graphite (2:1) after (a) 5 min, (b) 1 hour, (c) 5 hours, and (d) 24 hours hydrogenation reaction. Temperature 180 °C; pressure 14.7 Psig.

DISCUSSION

As can be deduced from the PCI profiles, the lower is the plateau pressure, the higher is the reactivity of the material [7]. The increase in reaction temperature from 150°C to 180°C brings the decrease in plateau pressure from 110 to 69 psig and the increase in the hydrogenation degree (H/Li molar ratio) from 0.78 to 0.81. However, further increasing temperature to 210°C, no plateau pressure was observed and the maximum H/Li molar ratio was only 0.11 (figure 2). It is quite interesting that higher temperature has negative effect on the hydrogenation of Li metal. We tentatively ascribe this deterioration due to: (1) the decomposition of active phase (GICs); (2) the complete coverage of active sites (out surface of GICs) by molten Li metal and thus the reaction routes for H_2 adsorption and activation are blocked; and (3) the deactivation of active sites for hydrogen absorption and activation/dissociation. XRD measurements showed that LiC_6 and LiC_{12} still existed in Li-graphite (10:1) sample after PCI measurement at 210°C (not shown) revealing the active phase is stable at this temperature. Considering that no H_2 can be absorbed by lithium metal at 210°C (figure 1) and the melting point of lithium is 185°C and the major component of the sample is lithium in Li-graphite (10:1), the coverage of GICs with lithium is possible. Furthermore, at higher temperatures, where the reaction between hydrogen and Li in GICs is kinetically favourable, there is a possibility that the activated hydrogen strongly bonds to the outmost surface of GICs thus the active sites for the reaction is deteriorated.

The hydrogenation reactivity of Li-C samples was also affected by the content of carbon. With the increase of Li:C molar ratio from 10:1 to 5:1 and to 2:1, the plateau pressure dropped from 69 to 19 and to 12 psig and the maximum H/Li molar ratio increased from 0.81 to 0.90 and to 0.96, respectively (Table 1). These results envisioned that higher content of carbon increased the hydrogenation reactivity and degree. This is reasonable for more GICs, which may serve as active centers for hydrogen adsorption and activation, presented in the lithium metal will provide more active sites and effective contact with lithium metal. As a compromise, the gravimetric percentage of H_2 uptake decreased from 9.9 to 7.0 and to 4.9 wt% when the Li:C molar ratio was changed from 10:1 to 5:1 and to 2:1 (Table 1).

CNTs used in this experiments are typically 20-40 nm in diameter and ~ 10 μm in length. Its (002) diffraction peak situates at $2\theta = 26°$. When mixed with lithium, the sample turned into golden-yellow colour and the main peak center shifted to $2\theta = 25°$. The peak shift reveals the expansion of interlayer distance which resulted from the intercalation of lithium. Comparing with Li-graphite, Li-CNTs shows higher H/Li molar ratio throughout the reaction temperature (figure 1), indicating Li-CNTs has higher hydrogenation reactivity than Li-graphite. Same results could also be observed from PCI results. The plateau pressure of Li-CNTs (30 psig) is about half of that of Li-graphite (69 psig) indicating that the former can adsorb H_2 easier (figure 4). Although CNTs possesses similar structure as graphite (4 μm in diameter and 7.9 $m^2 g^{-1}$ of specific surface area), it has much smaller size (20-40 nm in diameter) and high specific area (130 m^2/g). When added into lithium, more active sites should be developed. Thus, we observed quicker hydrogenation rate and lower plateau pressure. Also, the difference in the carbon texture and the types of Li-Cx may affect the hydrogenation reactivity.

It is well known that lithium has to be heated up to 600°C and keep at 147 psig of H_2 pressure for the synthesis of LiH [7]. Our PCI measurement confirmed these results (figure 1a). Hydrogen uptake by lithium metal started at 500°C and became vigorous above 650°C. However, according to reaction (1), the hydrogenation of Li is thermal-dynamically favourable. The strict experimental conditions suggest that high energy barrier, probably for the adsorption and activation/dissociation of hydrogen molecular, blocks the process of reaction. The addition of carbon into lithium metal undoubtfully promotes the hydrogenation capability of lithium metal and greatly reduces the reaction temperature to ~50°C, which is ~ 450°C lower than that for pure lithium metal. Carbon alone does not adsorb hydrogen under these experimental conditions. The great decrease in reaction temperatures should be attributed to the interaction between Li and carbon. Indeed, after hammering lithium and graphite for a period of time, a golden-yellow coloured structure was formed [1]. XRD characterization also demonstrated the existence of LiC_6 and LiC_{12} in the freshly made Li-C samples. Theoretical calculation has demonstrated that as Li intercalated with graphite, the Li 2s band would hybridize with C π bond, resulting in the formation of a new metal-like band. This new band may feed to the anti-bonding orbital of H_2 molecule and cause the weakening or even breaking the H-H bond [9].

The hydrogen uptake mechanism of Li-C can be understood as follows (figure 5): Li-C compounds serve as active centers for H_2 adsorption and activation/dissociation. After activation, the H atoms may diffuse and combine with lithium atoms nearby to form LiH. As revealed from in-situ XRD patterns (figure 4), the intensity of the metal lithium peak ($2\theta = 35.9°$) decreases 87%, while that of LiC_6 ($2\theta = 23.8°$) decreases only 10% after 1 h of reaction. This result indicated that at the early stage of the reaction, it is metal lithium rather than Li-C compounds that is responsible for the increase of LiH. After most of metal lithium was depleted, the lithium within the graphene layers could be drawn out gradually and graphite phase resumed simultaneously. Finally, all of lithium converted into LiH and graphite regenerated (figure 5c and 5d).

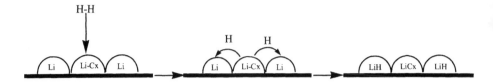

Figure 5. Schematic illustration for the hydrogenation of Li-C complex.

CONCLUSION

The addition of carbon materials such as graphite and CNTs into lithium metal greatly promotes its hydrogenation reactivity. The threshold temperature for hydrogenation was reduced from 500°C for lithium metal to 50°C for Li-C samples. At 180°C, ~10 wt% of hydrogen can be absorbed. *In-situ* XRD measurements demonstrated that the carbon could interact with lithium and form lithium carbon intercalation compounds such as LiC_6 and LiC_{12}. These compounds may severe as active centres for hydrogen adsorption and activation and should be responsible for the increase of reactivity. Increase of carbon content in Li-C system will enhance the hydrogenation reactivity. For the same content of carbon, Li-CNTs shows higher hydrogenation reactivity than Li-graphite.

REFERENCES

1. Z.S. Wronski, *International Materials Reviews*, **46**(1), 1 (2001).
2. A.C. Dollin, K.M. Jones, *Nature*, **386**, 377 (1997)
3. C. Liu, Y.Y. Fan, H.T. Cong, H.M. Cheng, M.S. Dresselhaus, *Science*, **286**, 1227 (1999).
4. P. Chen, X. Wu, J. Lin, K.L. Tan, *Science*, **285**, 91 (1999)
5. J.Z. Luo, L.Z. Gao, Y.L. Leung, and C.T. Au, *Catal. Lett.*, **66**, 91 (2000).
6. Powder Diffraction File [TM], volume released 1999.
7. M.V.C. Sastri, B. Viswanathan, and S. Srinivasa Murthy. Metal hydrides: fundamentals and applications, (Berlin: Springer-Verlag 1998).
8. Georg Brauer, Handbook of Preparative Inorganic Chemistry, Academic Press, New York London.
9. N.A.W. Holzwarth, S. Rabii, L.A. Girifalco, *Phys. Rev. B*, **18**, 5190 (1978).

Materials for Solar Energy

Mat. Res. Soc. Symp. Proc. Vol. 730 © 2002 Materials Research Society

Synthesis and Application of Electronic Oxides for Solar Energy

James M. Kestner, Anna Chorney, Joshua J. Robbins, Yen-jung Huang, Tyrone L. Vincent[1], and Colin A. Wolden
Department of Chemical Engineering and Division of Engineering [1]
Colorado School of Mines, Golden, CO 80401-1887, USA.
Lawrence M. Woods
ITN Energy Systems, Inc., Littleton, CO 80545, USA

ABSTRACT

Plasma-enhanced chemical vapor deposition (PECVD) is being developed as a flexible coating technology for a variety of oxides. In this paper we discuss the synthesis of transparent conducting oxides (TCOs), insulating oxides and electrochromic oxides. Tin oxide was synthesized using mixtures of $SnCl_4$ and O_2. By proper control of processing conditions the resistivity of this material may be varied from $10^{-3} < \rho < 10^5$ Ω-cm. Films of varying resistivity were employed as buffer layers in CdS/CdTe solar cells. Preliminary device results have demonstrated that integration of a tin oxide buffer layer was very beneficial for cell performance. In addition, we demonstrate the PECVD synthesis of WO_3 from $WF_6/O_2/H_2/Ar$ mixtures. The plasma process space that yielded adherent, transparent tungsten oxide was established. The deposited films were both amorphous and reversibly electrochromic. High temperature annealing above 400 °C converted the films into a polycrystalline state.

INTRODUCTION

Plasma-enhanced chemical vapor deposition (PECVD) is a flexible coating technology that has been employed for the fabrication of a variety of thin films. In this paper we discuss the synthesis of transparent conducting oxides (TCOs), insulating oxides and electrochromic oxides. TCOs form the transparent electrodes in photovoltaic (PV) devices, light emitting diodes (LEDs), and flat panel displays. Insulating oxides find application both as buffer layers and as protective coatings. Electrochromic materials may be reversibly converted from transparent to opaque to modulate sunlight in architectural applications. To date these materials have been fabricated primarily by physical vapor deposition techniques such as sputtering or laser ablation. In such cases the ceramic target sets the composition of the deposited film. The fabrication of multi-layer structures requires multiple targets, and it is difficult to make fine adjustments to film composition. Other drawbacks of PVD techniques include low target utilization and the production of asperities that compromise film smoothness.

We have been developing PECVD as an alternative oxide coating technology. Potential advantages include the ability to finely tailor film properties and produce functionally graded films by manipulation of process parameters. In addition multi-layer structures could be synthesized by simply altering the gas composition. In previous work we have discussed the use of PECVD to produce conducting tin oxide [1-2]. In this paper we discuss the application of these materials to the solar energy devices shown in Figure 1. The first device is the CdTe/CdS heterojunction solar cell shown in Figure 1(a). Cadmium telluride has a nearly ideal bandgap (~1.45 eV) for terrestrial photovoltaics. Highest efficiencies are obtained when light absorption

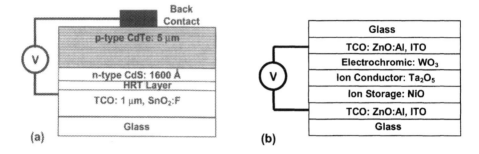

Figure 1: Devices of interest. (a) CdTe/CdS solar cell and (b) and a solid oxide electrochromic window.

in the cadmium sulfide window is minimized. However as the CdS layer is thinned, the probability of pinhole formation increases, causing localized CdTe/TCO junctions and inferior device performance. To mitigate this effect while maintaining thin CdS groups have attempted to insert high resistance transparent (HRT) buffer layers between the conducting TCO and the CdS layer [3]. Materials that have been employed include zinc stannate [3] and intrinsic tin oxide [4]. In this work we examine the use of PECVD SnO_2 of varying conductivity as the HRT layer.

The second device shown in Figure 1(b) is a solid oxide electrochromic window [5]. Electrochromic windows, with their ability to be darkened or lightened electronically, have the potential of reducing the annual U.S. energy consumption by several quadrillion (10^{15}) Btus. As one can see this device is composed of a stack of thin oxide films. Electrochromic materials such as tungsten oxide reversibly alter their optical properties in response to an electric field. The phenomenon is associated with the process of intercalation, the simultaneous transport of ions/electrons into and out of the film in response to an applied voltage. The intercalation reaction may be expressed as:

$$\underset{\text{(clear)}}{WO_3} + xM^+ + xe^- \leftrightarrow \underset{\text{(opaque)}}{M_xWO_3} \tag{1}$$

Ideally the device shown in Figure 1(b) could be synthesized in one step though PECVD by appropriately altering the gas-phase reactants. Having established TCO synthesis by PECVD, the next step is fabricating the electrochromic WO_3 layer. Here we investigate the process space for WO_3 deposition, test electrochromic behavior, and survey the thermal stability of the material.

EXPERIMENTAL DETAILS

Oxide films were deposited from various mixtures in a custom-made, stainless steel thin film deposition system that has been described previously [1]. Two-inch square glass substrates were clamped to the grounded electrode, which was maintained at the desired deposition temperature ± 1 °C by a PID-controlled electrical heater mounted on the back of the substrate assembly. Substrates were also placed on the powered electrode. Electronic mass flow controllers and an automated butterfly valve were employed to control composition and pressure. Thickness, transparency, and resistivity measurements were performed on all films. A Rigaku XRD system, with a Cu K_α anode, was used for the structural analysis. Typical 2θ scans in the range of 20° to 70° were made with a step size of 0.05° at an interval of 2 seconds. A Digital Instruments Nanoscope III was employed for AFM images. PV performance was measured under global AM 1.5 illumination at the Colorado School of Mines.

RESULTS & DISCUSSION

HRT Buffer layers in CdTe/CdS cells

Photovoltaic devices resembling those shown in Figure 1(a) were fabricated as described in details previously [6]. The base case substrates were commercial Libby-Owens-Ford sodalime glass slides coated with SnO_2:F as the TCO layer. Three additional substrates had a PECVD HRT layer deposited on the commercial TCO prior to CdS deposition. The HRT layer was composed of ~600 Å thick intrinsic tin oxide, with as-deposited resistivity varying between 100 – 1500 Ω-cm. The despoition conditions and properties are summarized in Table 1. It is known that thermal annealing can alter the conductivity of this material [7], and during cell processing the substrate is exposed to temperatures ranging from 400 - 600 °C.

Table 1. Processing conditions and properties of HRT layers.

Sample Letter	ρ [Ω-cm]	SnCl$_4$ [sccm]	O$_2$:SnCl$_4$ Ratio	Power [W]	Pressure [mtorr]	Substrate Temp [°C]	Rate [Å/min]
A	3	2.13	2.75	75	200	125	181
B	80	5.20	1.50	25	400	160	322
C	6000	1.20	1.50	125	200	75	180

The 1600 Å CdS layer was then deposited by chemical bath deposition, and the CdTe was deposited by atmospheric pressure chemical vapor deposition. The CdTe was subjected to post deposition thermal CdCl$_2$ treatment [8]. Numerous devices were fabricated from each substrate by evaporating gold contacts (0.03 cm^2) using a shadow mask. Indium solder was used to contact the TCO. With the exception of the presence of the HRT layer, all other cell processing was performed simultaneously.

The results of the study are shown below in Figure 2. Figure 2(a) compares the I-V curves of devices fabricated with and without an HRT layer. The devices incorporating the HRT layer show good rectifying behavior, while there is substantial roll off in the I-V curve from devices fabricated without HRT. Device efficiencies are compared in Figure 2(b). For the base case samples without a HRT layer, the measured efficiencies were between 7.5 and 8% as indicated by the horizontal line in Figure 2(b). In all cases the use of an HRT layer proved beneficial. The moderately resistive HRT layer, ρ = 45 Ω-cm, delivered an average device efficiency of 13.8% with the best device measuring 14.6%, an 83% increase over the base case values. For reference, the resistivity of the commercial TCO is ~5 x 10^{-4} Ω-cm. Device improvement was also seen for the HRT samples at 3 and 5000 Ω-cm, with typical values for both samples ranging from 10 – 11 %.

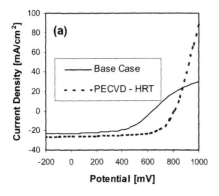 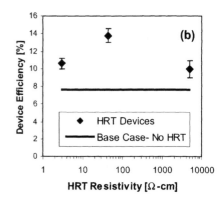

Figure 2: (a) Representative I-V curves obtained under global AM1.5 illumination; and (b) comparison of device efficiency as a function of HRT resistivity.

One contribution to the improved efficiency might be surface roughness with less sharp peaks. Figure 3 compares AFM images of the as-received TCO and the same material coated with PECVD tin oxide. The commercial TCO is produced by atmospheric chemical vapor deposition at high temperatures and is quite rough. The RMS roughness of the film shown in Figure 3(a) is ~400 Å. The roughness of the PECVD-coated sample shown in Figure 3(b) was reduced by half. Since the original roughness is similar in magnitude to the thickness of the CdS layer, the improvements may be in part related to improving the integrity of this window layer. However, the strong dependence on resistivity shown in Figure 2 indicates that electrical properties are critical as well. Further work is underway to better understand the role of HRT thickness and resistivity on device performance.

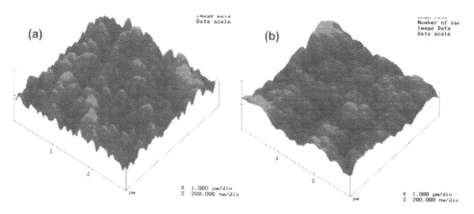

Figure 3: AFM images of (a) the as-received commercial SnO_2:F film and (b) a commercial film coated with HRT tin oxide deposited by PECVD. Scale is the same in both.

Deposition and Characterization of Electrochromic Tungsten Oxide

In previous studies of PECVD synthesis of tungsten oxide it has been shown that the ratios of WF_6: O_2: H_2 must be carefully controlled in order to produce quality films [9-10]. It was found in our system that the deposition rate was controlled primarily by WF_6 flowrate and argon dilution. For all experiments the WF_6 flowrate was set at 0.3 sccm and the argon flowrate was adjusted to maintain 90% dilution. The pressure, power and substrate temperature were all maintained at P = 250 mtorr; rf = 100 Watts, and T_s = 25 °C, respectively. Under these conditions the deposition rate was 150 – 200 Å/min. The use of higher WF_6 flowrates/lower dilution was prevented by the formation of significant amount of powder in the chamber. It was found that a O_2:WF_6 ratio between 3 and 6 yielded high quality, well-adhered films. Similarly, it was found that the H_2:WF_6 ratio must exceed 1.7. Films deposited outside of these parameters were either opaque or they delaminated easily from the glass substrates. The films' electrochromic nature was verified using the procedure developed by Crandall and Faughnan [11]. For this evaluation approximately 0.5 microns of tungsten oxide was deposited on commercial TCO-coated glass. The substrate was placed in a beaker containing dilute sulfuric acid. A DC voltage was applied between the TCO film and an electrode suspended in the solution. Rapid, reversible electrochromic behavior was observed by reversing the polarity.

The thermal stability and structure of these films were studied by annealing and X-ray diffraction, respectively. Figure 4 displays the diffraction patterns of an as-deposited film, and the same film annealed in argon at 465 °C. The as-deposited material was clearly amorphous. In addition, it was found that annealing at temperatures < 400 °C yielded similar XRD patterns to the as-deposited material. At even higher temperatures (465 °C) the amorphous material became crystalline, with the (200), (220) and (420) planes of WO_3 clearly identified. Understanding the dynamics of this phase transition is important to the synthesis of a device such as that described in Figure 1(b). It is well known that amorphous tungsten oxide is superior to crystalline material in terms of electrochromic response, so maintaining the disordered phase is critical. From our previous work [1], [2] it was identified that high quality TCO films required substrate temperatures of 250 – 350 °C in order to produce high conductivity. This requirement does not appear to be a limitation at this time. Future work involves studying the kinetics of intercalation using impedance spectroscopy coupled with optical transmittance. More work is required to explore the deposition process space as well.

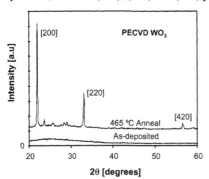

Figure 4: XRD pattern of as-deposited and annealed tungsten oxide films.

CONCLUSIONS

PECVD offers a promising alternative for the production of oxides integral to solar energy devices. PECVD has been used to synthesize conducting TCOs, intrinsic buffer layers, and electrochromic tungsten oxide. Intrinsic tin oxide deposited from $SnCl_4/O_2$ mixtures was employed as a buffer layer in the fabrication of CdTe/CdS solar cells. Dramatic increases in device efficiency were observed that depended on the resistivity of the buffer layer. The process space for the synthesis of transparent, adherent tungsten oxide was identified for $WF_6/O_2/H_2/Ar$ mixtures. The films were reversibly electrochromic. An XRD analysis showed that the as–deposited films were amorphous, and converted to polycrystalline at temperatures greater than 400 °C.

ACKNOWLEDGEMENTS

This work was supported in part by the National Science Foundation through grants #DMII-9978676, CTS-0093611, and through the National Renewable Energy Laboratory PV Partnership Program subcontract ZAK-8-17619-03.

REFERENCES

1. J. J. Robbins, R. T. Alexander, M. Bai, Y.-J. Huang, T. L. Vincent, and C. A. Wolden, *J. Vac. Sci. Technol. A*, **19** 2762 (2001).
2. J. J. Robbins, R. T. Alexander, W. Xiao, T. L. Vincent and C. A. Wolden, *Thin Solid Films*, (2002) (in press).
3. X. Wu, X., P. Sheldon, Y. Mahathongdy, R. Ribelin, A. Mason, H. R. Moutinho, and T. J. Coutts, in *NCPV Photovoltaics Program Review*, edited by M. Al-Jassim, J. P. Thorton, and J. M. Gee, (American Institute of Physics CP462, Woodbury, NY, 1999) pp. 37-41.
4. X. Li, R. Ribelin, Y. Mahathongdy, D. Albin, R. Dhere, D. Rose, S. Asher, H. Moutinho, and P. Sheldon, in *NCPV Photovoltaics Program Review*, edited by M. Al-Jassim, J. P. Thorton, and J. M. Gee, (American Institute of Physics CP462, Woodbury, NY, 1999) pp. 230-235.
5. J. G. H. Mathew, S.P.S., M. J. Cumbo, N. A. O'Brien, R. B. Sargent, V. P. Raksha, R. B. Lahaderne, and B. P. Hichwa, *J. Non-Cryst. Solids,* **218** 342 (1997).
6. J. M. Kestner, C. A. Wolden, P. V. Meyers, L. L. Raja, and R. J. Kee, in *Proceedings of the 28th IEEE Photovoltaic Specialists Conference* (Piscataway, NJ, 2000) pp. 595-598.
7. J. J. Robbins, Y.-J. Huang, M. Bai, T. L. Vincent, and C. A. Wolden, *Mater. Res. Soc. Symp. Proc.* **666** F1.7.1 (2001).
8. Y. Mahathongdy, D. S. Albin, C. A. Wolden, and R. B. Baldwin, in *NCPV Photovoltaics Program Review*, edited by M. Al-Jassim, J. P. Thorton, and J. M. Gee, (American Institute of Physics CP462, Woodbury, NY, 1999) pp. 236-241.
9. C. E. Tracy and D. K. Benson, *J. Vac. Sci. Technol. A* **4** 2377 (1986).
10. W. B. Henley and G. J. Sacks, *J. Electrochem. Soc.* **144** 1045 (1997).
11. R. S. Crandall and B. W. Faughan, *Appl. Phys. Lett.* **26** 120 (1975).

Mat. Res. Soc. Symp. Proc. Vol. 730 © 2002 Materials Research Society   V3.4

Fabrication of Copper-Indium-Disulfide Films onto Mo/Glass Substrates Using Pulsed Laser Deposition

R. Mu, M.H. Wu, Y. C. Liu[1], A. Ueda, D.O. Henderson, A.B. Hmelo[2] L.C. Feldman[2] and A. Hepp[3]

Chemical Physics Laboratory, Department of Physics, Fisk University, Nashville TN 37208, USA
[1]Open Laboratory of Excited State Processes, Changchun Institute of Optics, Fine Mechanism and Physics, Chinese Academy of Sciences, Changchun 130021, People's Republic of China.
[2]Department of Physics and Astronomy, Vanderbilt University, Nashville, TN 37235, USA
[3]NASA Glen Research Center, Cleveland, OH 44135, USA

ABSTRACT

Pico-second pulsed laser deposition (PLD) was employed to fabricate copper indium disulfide (CIS) thin films onto pure silica and Mo coated glass substrates. By properly preparing the target materials and controlling the elemental ratio of the Cu, In and S in the deposited film followed by post-thermal annealing, good quality copper-indium-disulfide(CIS) films can be obtained. A series of characterizations were conducted including XRD, RBS, IR, UV-Vis, AFM and STM analyses.

INTRODUCTION

Photovoltaics (PV), as an alternative energy source, can be used for anything that requires electricity ranging from small and remote applications to central power stations. PVs have great economic and environmental benefits and are versatile as an energy source.

Thin film photovoltaic devices offer several advantages over other solar cells. Among these benefits are possible lower cost, a large-scale application, few limitations of shapes and configurations, light weight, and high radiation resistance. These factors, in particular, the light weight and high radiation resistance make these materials attractive for space applications.

Chalcopyrite semiconductors, such as $CuInS_2$ and $CuIn_xGa_{1-x}Se_2$, are very promising materials as active layers for solar cells [1-7]. They not only have the optical band gap close to the peak of the solar energy spectrum, but also have a high absorption cross section above 10^4 cm^{-1}. Thus, a $1 - 2$ μm thick film is sufficient to absorb over 99% incident solar energy above the band gap [1]. It was reported recently that the conversion efficiency has reached up to 18.8% at the laboratory level for $CuIn_xGa_{1-x}Se_2$ at the National Renewable Energy Laboratory (NREL).

On the other hand, the current achievable energy conversion efficiency is not limited by fundamental physics of particular materials. Rather, it is limited by fabrication techniques, film quality control, and tailoring of the optical band gap, carrier types and carrier concentration via control of the intrinsic stoichiometry of the film and phase compositions.

Pulsed laser deposition (PLD) has been used successfully to grow many different types of low dimensional materials including high quality films and nanocrystals. It is highly compatible with other high vacuum techniques including MBE, sputtering, e-beam evaporation and so on. Single-crystal semiconductor epitaxial thin films can be fabricated via laser-MBE [2]. Superlattice and multilayer structures can also be easily made by using multiple targets at a relatively low temperature [3].

In this paper we report the application of PLD to 1) obtain high quality CIS films on various substrates by controlling deposition parameters, such as substrate temperature, laser energy density, substrate-to-target distance, types of backing gas and gas pressure, as well as the chemical composition and structure of the targets; 2) establish an optimized post-thermal annealing procedure to enhance the film and device quality; and 3) develop a baseline characterization procedures for evaluating CIS thin film photovoltaic materials by conducting a comprehensive study of the optical, electronic, and structural properties of CIS thin film.

EXPERIMENTAL

The second harmonic of a picosecond Nd:YAG laser (model PY61c-10, Continuum) at 532 nm was used for PLD experiments. The repetition rate was 10 Hz. A focused beam was directed onto the target with an irradiated spot size of ~ 0.5 mm. The pulse energy was ~ 2 mJ per pulse and the pulse-to-pulse energy fluctuation was < 10%. The vacuum chamber was pumped down to ~ 2×10^{-6} torr, and then was back-filled with 99.998% Ar gas to a 1 -50 m torr range. The chamber was then isolated from the pumping station to ensure that the volatile material produced from laser ablation of the target remained in the chamber. A typical deposition time required 30 minutes. An increase of the chamber pressure of < 2 m torr was noted during the PLD and was attributed to the formation of volatile species generated from the ablation of the target material (sulfur containing compounds).

Two experimental procedures were used to prepare the ablation targets. In procedure 1, we used ultra-pure In_2S_3, Cu_2S and S as the starting materials. The three compounds were mixed to render copper rich, indium rich, sulfur rich, and stoichimetric mixtures. The targets were made by pressing the compound mixture into pellets after grinding in a mortar and vibratory ball mill. The pellets were subsequently annealed in an Ar atmosphere for 2 hrs at 250°C. In procedure 2, the target material was prepared by mixing In_2S_3, Cu_2S and S to obtain the stoichiometry of In:Cu:S = 1:1.25:2.4. The mixture then was sealed in a quartz ampoule under the vacuum (10^{-5} torr). Then, the ampoule was slowly heated to 400°C for 48 hrs. Thereafter, the ampoule was further heated to 500°C for 12 hrs. The ampoule was slowly cooled down to room temperature. Pellets were made by pressing the processed powder from the ampoule. The pellets were then used as the target materials.

The CIS film made by PLD was then thermally annealed in a quartz ampoule. The film deposited on Mo/glass substrate was sealed in 99.998% pure Ar gas with a pressure of slightly above 1 atm. The annealing temperatures used were 350, 400, 450, 500 and 550°C. The uncertainty of the annealing temperature was estimated to be less than 25°C. Each sample was annealed at a pre-set temperature for 4 hrs before any further analyses were done. No serial annealing was performed on any sample to ensure that the films did not carry history from any previous thermal treatment.

The elemental composition analysis of CIS films was done by Rutherford backscattering spectrometry (RBS) with 1.8 MeV alpha particles from a Van der Graaff accelerator. The number of scattered ions were integrated and their energies were measured with a silicon surface barrier detector.

X-ray diffraction (XRD) spectra were obtained using a Scintag X1 Advanced Diffraction System equipped with a solid-state Peltier detector, a high-temperature unit integrated with a sample holder, Micristar temperature control unit, standard room-temperature sample holder, and a computer controlled gas delivery system. A copper x-ray tube was used as the source of radiation. All XRD spectra were collected at room temperature.

A Hitachi UV-Vis-NIR 3501 spectrophotometer was used to study the optical properties of the CIS film. In order to eliminate any possible scattering losses due to the surface roughness and grain boundaries, an integrating sphere accessory was used to collect the transmission spectra. A conventional simple transmission measurement was also conducted to compare with the spectra obtained with the integrating sphere. The result indicates that the scattering loss was negligible. However, all the optical spectra reported here were absorbance spectra. Reflectance measurements at 5° incidence were also carried out using a reflectance accessory with a standard mirror as the reference.

Infrared transmission and reflectance measurements were made with a bench-top Bomem MB102 spectrometer at 4 cm^{-1} resolution.

A TappingMode atomic force microscope (TM-AFM), Nanoscope III from DI, was used to study 1) the surface roughness of as-deposited CIS films and 2) thermal annealing effects on the morphology of the film. An E-scanner was used with a maximum scan size of ~ 16 x 16 μm^2. The lateral resolution is estimated to be 10 - 30 nm depending on the particular tip used and the tip history. Typically, three tips were used alternately for one sample to eliminate image distortion due to any possible tip artifacts. The accuracy in height is ~1 nm or better depending on the scan size and the surface roughness. The scan rate was ~1.5 Hz. No filters were used to obtain images.

RESULTS AND DISCUSSION

As mentioned in the experimental section, we have devised two procedures to prepare the target materials with various different stoichiometries. However, the target prepared with the stoichiometry of In:Cu:S = 1:1.25:2.4 and annealed in an ampoule under a vacuum provided high quality CIS films compared to those prepared by procedure 1. Thus, only the results for procedure 2 are discussed in this paper.

Fig. 1 shows a RBS spectrum of a target prepared by procedure 2. The elemental composition of the target is In:Cu:S =1:1.25:2.4 with 3% uncertainty based on RUMP computer simulation. From the X-ray diffraction measurements, the observed XRD peaks fit well with CuInS$_2$ powder

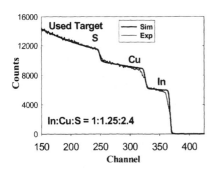

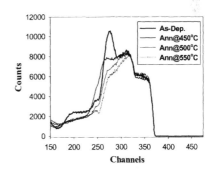

Figure 1 RBS Spectra of target used for fabricating CIS films on Mo/glass substrates (dashed line) along with the computer simulated spectrum (solid line).

Figure 2 RBS spectra of CIS films deposited on a Mo/glass, as-deposited and annealed at 450, 500 and 550°C

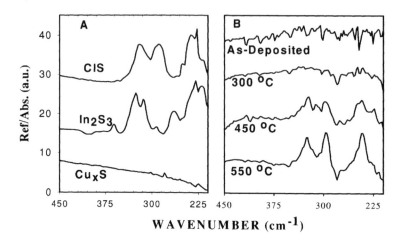

Figure 3 Infrared reflectance spectra of the CIS film on a Mo/glass substrate, as-deposited and annealed at 300, 450 and 550°C.

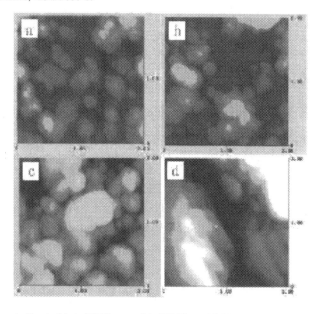

Figure 4 . TappingMode AFM images of the CIS film on Mo/glass, as-deposited and annealed at 450, 500 and 550°C.

diffraction lines suggesting the target is mainly polycrystalline $CuInS_2$ with a chalcopyrite structure. Fig. 2 shows RBS spectra of as-deposited film and the same film after thermal annealing at 450, 500 and 550°C for 4 hours under Ar. The results show that as the annealing temperature increases, the two peaks representing sulfur and copper in the film decrease indicating a loss of these two elements. But the amount of decrease in both elements is less than 5% when compared with as-implanted film. It is important to point out that the stoichiometry of the as-deposited film is In:Cu:S = 1:1.05:2.05 with 5% uncertainty due to the surface roughness

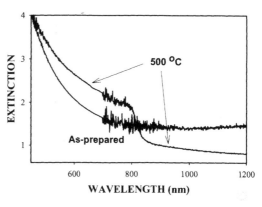

Figure 5 Optical absorption spectra of as-deposited and the thermally annealed CIS film at 500°C.

of the film. The XRD spectra show that for the as-deposited film, no definite XRD diffraction lines were observed. However, the samples annealed at the temperature above 450°C gave typical $CuInS_2$ chalcopyrite XRD lines. These spectral lines became narrower as the annealing temperature increased. Fig. 3 illustrates the infrared spectra of $CuInS_2$, In_2S_3 and Cu_xS with x = 2 as the spectral reference (3A) and the as-deposited and thermally annealed films (3B). For the as-deposited sample and the samples annealed below 300°C, no well-defined phonons were observed in 200 – 400 cm^{-1} region. When the annealing temperature was above 400°C, three phonons at 330, 290, and 242 cm^{-1} were observed indicating the formation of crystalline CIS [9]. Fig. 4(a) – 4(d) show the tapping mode AFM images of the samples before and after annealing at 450, 500 and 550°C. Clearly, when the annealing temperature is below 550°C, no obvious surface morphology change could be observed. When the temperature reached 550°C, there is a clear indication that the crystallization and re-growth occurred on the surface. In fact, micron-size, faceted CIS particles were clearly observed. Fig. 5 shows the UV-Vis spectra of as-deposited and annealed film at 500°C. In the case of the as-deposited film, a gradual optical absorption was observed starting below 750 nm. When the film was annealed at 500°C, a clear absorption onset was observed near 840 nm (1.475 eV) [10]. Based on the film thickness and optical absorption spectra, the amplitude of the absorption coefficient above the band edge exceeds 10^4 cm^{-1}.

It is clear that good quality CIS films can be obtained by PLD of a CIS target and post thermal annealing. Having developed PLD as synthetic approach for fabricating CIS films and an annealing protocol, it is of interest to compare the PLD approach and our annealing routine with earlier reports. As previously reported in literature, the thermal annealing temperature of CIS film was limited to < 450°C; high temperature annealing will lead to the deviation from the correct stoichiometry. As a result, other phases can be formed as we reported earlier [10-12]. However, our results show that the integrity of stoichiometry of the CIS films were maintained at temperatures >450° C. A plausible explanation for not observing other secondary phases at higher temperature may be due to the fact that the original as-deposited film in our case is Cu and S rich which compensates the loss of these two elements during annealing. The other reason is that the thermal annealing conditions used in our experiments differed from those reported in the literature. Specifically, we did not anneal the films under an inert gas flow nor under a dynamic vacuum as was the case previous reports in the literature. Consequently, any volatile matter, such as sulfur and/or sulfur related species produced during annealing, is retained in the chamber and served as a part of the annealing environment, whereas both gas purge and vacuum annealing conditions will carry away or pump out the volatile matter.

CONCLUSION

Pulsed laser deposition can be effectively used to fabricate multi-component semiconductor thin films. A good quality CIS film has been obtained by ablating slightly copper and sulfur rich target material followed by thermal annealing. It is shown that the annealing conditions such as temperature, environment and time are important factors to achieve a high quality CIS film on Mo/glass substrates.

ACKNOWLEDGMENTS

This work was supported under NASA Grant No. NCC3-575, Consortium for Advancement of Renewable Energy Technology and NASA supported Center for Photonic Materials and Devices.

REFERENCES

1. D.O. Henderson, R. Mu, et al., Materials and Design **22**, (2001) 585.
2. J. T. Cheung and J. Madden, J. Vac. Sci. Technol. **B 5** 705 (1987).
3. J. J. Dubowski, D. F. Williams, P. B. Sewell, and P. Norman, Appl. Phys. Lett. **46** 1081 (1985).
4. J. T. Chung and H. Sankur, CRC Cris. Rev. Solid State Mater. Sci. **15** 63 (1988).
5. J. Hedstrom, M. Bodegard, A. Kylner, L. Stolt, and H.W. Shock, *Proceedings of the 23rd Institute of Electrical and Electronic Engineers Photovoltaic Conference*, Lewisville, Kentucky, pp. 364 (1993).
6. J. Ermer, R. Gay, D. Pier, and D. Tarrant, J. Vac. Sci. **A 11**, (1993) 1888.
7. S.P. Grindle, A.H. Clark, S. Rezaie-Serej, E. Falconer, J. McNaily, and L.L. Kazmerski, J. Appl. Phys. **51** (1980) 5464.
8. D. Schmid, M. Ruckh, F. Grunwals, and H.W. Schock, J. Appl. Phys. **73** (1993) 2902
9. G. Morell, R.S. Katiyar, S.Z. Weisz, T. Walter, H.W. Schock, and I. Balberg, Appl. Phys. Lett. **69**, (1996) 987.
10. L. Stolt, J. Hedstrom, J. Kessler, M. Ruckh, K.O. Velthaus, and H.W. Schock, Appl. Phys. Lett. **62** (1993) 597.
11. H. Xiao, L.C. Yang, and A. Rockett, J. Appl. Phys. **76** (1994) 1503.
12. A. Rockett and R.W. Birkmire, J. Appl. Phys. **70** (1991) R81.

Mat. Res. Soc. Symp. Proc. Vol. 730 © 2002 Materials Research Society

Metal-Organic Chemical Vapour Deposition of II-VI Semiconductor Thin Films Using Single-Source Approach

Mohammad Afzaal, David Crouch, Paul O'Brien and Jin-Ho Park

The Manchester Materials Science Centre and Department of Chemistry, University of Manchester, Oxford Road, Manchester, M13 9PL, UK.
E-mail: mohammad.afzaal@stud.man.ac.uk; paul.obrien@man.ac.uk; jin-ho-park@man.ac.uk;

ABSTRACT

Thin films of CdS and CdSe have been deposited on glass substrates by low pressure metal-organic chemical vapour deposition (LP-MOCVD) using $Cd[(EP^iPr_2)_2N]_2$ (E = S, Se) as single-source precursors. These air-stable precursors are volatile, making them suitable for the deposition of thin films. As-deposited films were crystalline metal chalcogenides, as confirmed by X-ray powder diffraction (XRD), and their morphologies were studied by scanning electron microscopy (SEM).

INTRODUCTION

The direct transition nature of II/VI materials makes them suitable for use in optoelectronic devices, e.g. cadmium/zinc chalcogenides are useful materials for solid-state solar cells,[1] photoconductors, field effect transistors, sensors and tranducers.[2]

Research on single-source precursors for deposition of II/VI thin films has been very active, as they have several advantages over conventional MOCVD using dual source precursors including lower growth temperatures, no premature reactions and good quality homogenous films. Bochmann *et al.* have produced a range of precursors based on 2,4,6,-tri-*tert*-butylphenylchalcogenolate which have been used to deposit thin films of the metal sulfides/selenides in preliminary low pressure growth experiments.[1-3] In related work, Arnold and coworkers have also deposited a range of metal chalcogenides using bulky silicon systems (e.g. $[MESi(SiCH_3)_3]$ M = Zn (II), Cd (II), Hg (II) and E = S, Se or Te).[4,5] Other classes of molecules, which have proved useful for the deposition of thin films, include coordination complexes such as dialkyldichalcogenocarbamates or dithiophosphinates.[6-10]

We have previously identified a new class of organometalic single-source precursors based on $[NH(SePPh_2)_2]$ ligands (an analogue of the β–diketonates), $M[(SePPh_2)_2N]_2$ [(M = Cd (II) or Zn (II)] complexes and have been recognised as suitable precursors for zinc/cadmium selenide films by LP-MOCVD.[11] In this paper, we focus on deposition and characterisation of CdS and CdSe films grown by low pressure MOCVD on glass substrates by using alkyl substituted $Cd[(EP^iPr_2)_2N]_2$ (E = S, Se).

EXPERIMENTAL DETAILS

Precursor Synthesis

$^iPr_2P(Se)NHP(Se)^iPr_2$ (1): A solution of diisopropylchlorophosphine (72 g , 0.472 mol) in anhydrous toluene (100 cm^3) was added dropwise over 45 minutes to a stirred solution of 1,1,1,3,3,3-hexamethyldisilazane (38.07 g , 0.236 mol) in anhydrous toluene (150 cm^3) at 60 °C.

The reaction was stirred at 60-70 °C for 4 hours and cooled to room temperature. Selenium powder (37.27 g , 0.472 mol) was added and the suspension heated at reflux for 12 hours. The resulting orange homogeneous solution was concentrated to ca 100 cm^3 and cooled at O °C for 12 hours. The resulting white precipitate was recovered and washed with diethyl ether (100 cm^3) and cold toluene (100 cm^3). Recrystallisation from chloroform/hexane, yield 71%, off white powder. (Mpt 171-173 °C); FT-IR (fourier transform-infrared): 3209 (v N-H), 1385 (δ N-H), 907, 879 (v P-N-P), 489 (v P-Se) cm^{-1}; ^{1}H NMR (nuclear magnetic resonance) (CDCl$_3$): δ = 1.3 (m, 24H, 8CH$_3$-R), 2.75 (m, 4H, 4R-CH), 3.1 (s, 1H, N-H); ^{31}P NMR (CDCl$_3$): δ = 90.829, MS-FAB (mass spectroscopy-fast atom bombardment): m/z = 408 (M + H$^+$); Anal. Calcd. for C$_{12}$H$_{29}$NP$_2$Se$_2$: C, 35.39; H, 7.18; N, 3.44; P, 15.28 %. Found: C, 35.31; H, 7.26; N, 3.49; P, 15.10 %.

^{i}Pr$_2$P(S)NHP(S)^{i}Pr$_2$ (2): A solution of ^{i}Pr$_2$PNHP^{i}Pr$_2$ was prepared as for 1. Sulfur (15.14 g, 0.471 mol) was added and the reaction was refluxed for 12 hours and cooled to 0 °C. The resulting white precipitate was filtered and washed with CS$_2$ (30 cm^3) and petroleum ether (30 cm^3). Recrystallisation from dichloromethane/hexane, yield 61% of the pure compound. MS (FAB): m/z = 314 (M + H$^+$); Anal. Calcd for C$_{12}$H$_{29}$NP$_2$S$_2$: C, 46.00; H, 9.26; N, 4.47; P, 19.79 %. Found: C, 46.30; H, 9.66; N, 4.44; P, 20.14 %.

Cd[(SP^{i}Pr$_2$)$_2$N]$_2$ (3): Sodium methoxide (1.09 g, 19.7 mmol) was added to a stirred solution of 2 (6 g, 19.16 mmol) in anhydrous methonal (100 cm^3) and stirred for 10 minutes at room temperature. CdCl$_2$ (1.76 g, 9.56 mmol) was added to the resulting solution and white precipitate was formed immediately. The suspension was furthur stirred for 2-3 hours and it was filtered and dried under vaccum. Recrystallisation from dichloromethane/petroleum ether (40-65 °C). Yield 89% white colour product. FT-IR: 1224, 767 (v P-N-P); ^{1}H NMR (CDCl$_3$): δ = 1.3 (m, 48H, 16CH$_3$-R), 2.1 (m, 8H, 8R-CH); ^{31}P NMR (CDCl$_3$): δ = 62.59; MS (FAB): m/z = 878 (100%, M + H$^+$); Anal. Calcd for C$_{24}$H$_{56}$N$_2$P$_4$S$_4$Cd: C, 39.12; H, 7.60; N, 3.80; P, 16.81 %. Found: C, 39.30; H, 7.80; N, 3.74; P, 17.11 %.

Cd[(SeP^{i}Pr$_2$)$_2$N]$_2$ (4): Sodium methoxide (0.56 g, 9.82 mmol) was added to a stirred solution of 1 (4 g , 9.82 mmol) in anhydrous methanol (100cm^3). The resulting pink solution was stirred at room temperature for 10 minutes. Cadmium chloride (0.90 g, 4.91 mmol) was added yielding an immediate off white precipitate. The suspension was stirred at room temperature for 2 hours. The recovered solid was washed with methanol (100 cm^3) and dried under vacuum. Recrystallisation from chloroform/methanol. yield 99% white crystals. FT-IR: 1226, 763 (v P-N-P), 425 (v P-Se) cm^{-1}; ^{1}H NMR (CDCl$_3$): δ = 1.15 (m, 48H, 16CH$_3$-R), 2.1 (m, 8H, 8R-CH); ^{31}P NMR (CDCl$_3$): δ = 57.101. MS (FAB): m/z = 926 (M + H$^+$); Anal. Calcd for C$_{24}$H$_{56}$N$_2$P$_4$Se$_4$Cd: C, 31.16; H, 6.10; N, 3.03; P, 13.39 %. Found: C, 31.33; H, 6.15; N, 2.98; P, 13.53 %.

Deposition of films and characterizations

Low Pressure Metal-Organic Chemical Vapour Deposition (LP-MOCVD): Thin films of metal chalcogenides were grown on borosilicate glass in a low pressure ($\approx$ 10^{-2} Torr) MOCVD reactor tube which has been described elsewhere.[12] A graphite susceptor holds the substrate dimensions (10 mm x 15 mm) which was heated by a tungsten halogen lamp.

Film characterisations: X-ray Powder diffraction studies were performed on a Philips X'Pert MPD diffractometer using monochromated CuK$_\alpha$ radiation. The samples were mounted flat and scanned from 10° to 90° in a step size of 0.04° with a count rate of 2.5 sec. Samples were carbon coated using an Edward coating system model E306A before SEM and EDAX. SEM was carried out using Philips XL30 FEG and EDAX (energy dispersive X-ray analysis) was preformed on an EDAX DX4. TGA (thermogravimmetric analysis) was done on a Seiko model SSC/S200 with a heating rate of 10 °C/min under nitrogen. Electronic absorption spectra were recorded on Heλios-Beta Thermospectronic spectrophotometer.

RESULTS AND DISCUSSION

The syntheses of [NH(EP^{i}Pr$_2$)$_2$] (E = S, Se) ligands were done according to previously reported methods.[13,14] The procedure involves deprotonation of the N-H moiety, using sodium methoxide, to form anionic chelate complexes for subsequent reaction with CdCl$_2$. These reactions have many advantages over conventional routes using metal carbonates;[13] high yield is obtained particularly for selenium complexes and the reactions occur at room temperature in dry methanol.

The volatility characteristics of the precursors were determined by thermogravimetric analysis (TGA), showing clean sublimation without any residues, which is a desirable characteristic of precursors for MOCVD studies. Weight loss for compound **3** was found to be between *ca.* 250 – 365 °C and compound **4** was between *ca.* 340 – 425 °C. Deposition of thin films on glass substrates was attempted over a range of substrate temperatures, maintaining a precursor temperature of 325 °C for one hour, with *ca.* 200 mg of precursor used for each experiment.

Cadmium Sulfide

CdS films were grown from Cd[(SP^{i}Pr$_2$)$_2$N]$_2$ at growth temperatures of 425 and 450 °C. Deposited films were yellow, transparent and well-adherent to the glass surface (Scotch-tape test). At the highest temperature (475 °C), slightly dark yellow-brown film was observed after one hour of growth.

As-deposited cadmium sulfide films were analysed by XRD, which indicated that the hexagonal CdS [JCPDS (Joint Committee of Powder Diffraction Standard): 06-0314) has been deposited (Fig. 1). XRD patterns of deposited films grown at 425 and 450 °C show a preferred orientation along the (002) plane independent of growth temperature. However, film deposited at 475 °C was found to be amorphous. SEM studies indicate that the morphology of film deposited at 450 °C consists of randomly oriented domains of compacted thin acicular crystallites. SEM image of the cross-section of the film grown at 450 °C, shows that as-deposited film is approximately 1.75 µm thick (error value ≈ 5 %) after one hour growth (Fig. 2). Similar growth morphology was observed for the film deposited at 425 °C. On the basis of EDAX analysis, the film was found to be slightly cadmium rich (52 %) and also, 2 % of phosphorous was detected.

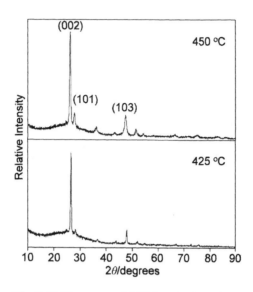

Fig. 1. XRD patterns of CdS deposited on glass.

Fig. 2. SEM micrographs of CdS deposited on glass at 450 °C.

Cadmium Selenide

The deposition of CdSe films was studied for substrate temperatures of 425-500 °C, with the volatisation temperature of the precursor set at 325 °C. Little or no deposition was observed at substrate temperatures of 425 and 450 °C. At growth temperatures of 475 and 500 °C, deposited films were black and adherent to the surface. XRD studies of the grown films indicated that polycrystalline hexagonal CdSe films (JCPDS: 08-0459) were deposited at 475 and 500 °C with a preferred orientation in the (100) plane (Fig. 3). Film deposited at 500 °C was slightly more ordered than the film grown at 475 °C due to the absence of (002) plane at 25.5° 2θ. In contrast, films grown from $Cd[(SePPh_2)_2N]_2$ by LP-MOCVD also resulted in hexagonal CdSe at 500 and 525 °C but show different preferred orientations [(002) *vs.* 101)] respectively at different growth temperatures.

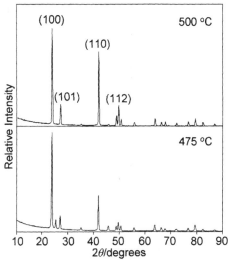

Fig. 3. XRD patterns of CdSe deposited on glass.

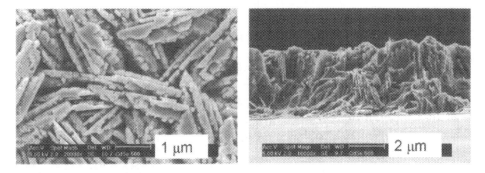

Fig. 4. SEM micrographs of CdSe deposited on glass at 500 °C.

SEM studies indicate that the morphology of the film grown at 500 °C consists of randomly oriented platelets (approximately 1 μm in size) giving a growth rate of *ca.* 3 μmh^{-1} (Fig. 4). The surface morphology can also be compared with previous attempts to deposit CdSe films from Cd[(SePPh$_2$)$_2$N]$_2$, which exhibited low degree of crystallinity hence indicating smaller particle size and growth rate was also significantly lower (1 μm h^{-1}).[17] EDAX analysis of the film deposited at 500 °C from Cd[(SeP^{i}Pr$_2$)$_2$N]$_2$ shows the presence of cadmium (54%) and selenium (46%). No phosphorous contamination was detected as an impurity.

The optical properties of the films were measured using UV/Vis Spectroscopy. Optical band gaps of films were calculated from electronic absorbance data by plotting α^2 vs. E, where α is the absorption coefficient and E is the photon energy. The bandgap values extrapolated for the CdS and CdSe thin films were 2.3 and 1.7 eV. The bandgap values are very close to reported values [(hexagonal CdS: 2.42 eV and hexagonal CdSe: 1.70 eV)].[15]

CONCLUSION

This study has shown high quality films of cadmium chalcogenides can be produced from the air stable precursors $Cd[(EP^iPr_2)_2N]_2$ (E = S, Se) by low-pressure metal-organic chemical vapour deposition. Another significant advantage is complete volatisation of the precursors, making them ideal for MOCVD studies. Analysis of the as-deposited films show that hexagonal phase of CdS/CdSe is prepared, regardless of difference in growth temperatures.

ACKNOWLEDGEMENTS

Authors thank the EPSRC, UK for the grants as a part of programme studying new precursors for chalcogenides. POB is Sumitomo/STS visiting Professor of Materials Chemistry at ICSTM.

REFERENCES

[1]   M. Bochmann, K. J. Webb, M. Harman and M. B. Hursthouse, *Angew. Chem, Int. Ed. Engl.* **638**, 29 (1990).
[2]   M. Bochmann, K. J. Webb, M. B. Hursthouse and M. Mazid, *J. Chem. Soc, Dalton Trans.* **9,** 2317 (1991).
[3]   M. Bochmann, and K. J. Webb *J. Chem. Soc, Dalton Trans.* **9**, 2325(1991).
[4]   B. O. Dabbousi, P. J. Bonasia and J. Arnold, *J. Am. Chem. Soc.* **113**, 3186 (1991).
[5]   P. J. Bonasia and J. Arnold, *Inorg. Chem.* **31**, 2508 (1992).
[6]   R. Nomura, T. Murai, T. Toyosaki and H. Matsuda, *Thin Solid Films.* **4,** *271* (1995).
[7]   D. M. Frigo, O. F. Z. Khan and P. O'Brien, *J. Cryst. Growth.* **96**, 989 (1989).
[8]   M. B. Hurthouse, M. A. Malik, M. Motevalli, and P. O'Brien, *Polyhedron.* **11**, 45 (1992).
[9]   Y. Takahashi, R. Yuki, M. Sugiura, S. Motijima and K. Sugiiyama, *J. Cryst. Growth.* **50**, 491 (1980).
[10]   M. Chunggaze, J. McAleese, P. O'Brien and D. J. Otway, *Chem. Commun.* **7**, 833 (1998).
[11]   M. Afzaal, D. Crouch, P. O'Brien, J. –H. Park and J. D. Woollins, *submitted to Advanced Materials, CVD.*
[12]   M. A. Malik and P. O'Brien, *Adv. Mater. Opt. Electron.* **3**, 171 (1994).
[13]   D. Cupertino, R. Keyte, A. M. Z. Slawin, D.J. Williams and D. J. Woollins, *Inorg. Chem.* **35**, 2695 (1996).
[14]   D. Cupertino, D. J. Birdsall, A. M. Z. Slawin and J. D. Woollins, *Inorg. Chem. Acta.* **290**, 1 (1999).
[15]   T. Trindade, P. O'Brien and N. L. Pickett, *Chem. Mater.* **13**, 3845 (2001).

Mat. Res. Soc. Symp. Proc. Vol. 730 © 2002 Materials Research Society

Organic-semiconductor-based all-solid-state photoelectrochemical cells

Robert Hudej, Egon Pavlica and Gvido Bratina
Nova Gorica Polytechnic, Vipavska 13 SI-5001 Nova Gorica, Slovenia
Urška Lavrencic- Štangar, Angela Šurca Vuk and Boris Orel
National Institute for Chemistry, Hajdrihova 19, SI-1000 Ljubljana, Slovenia

ABSTRACT

The solid-state solar cells comprising dye-sensitized nanostructured SnO and vacuum-evaporated 3,4,9,10-perylene tetracarboxylic dianhydride (PTCDA) layers exhibit significant photoresponse. Despite of unfavorable electronic energy level alignment at the dye/PTCDA interface, photon-to-electron conversion efficiencies as high as 1% were observed.

INTRODUCTION

Dye-sensitized photochemical solar-cells (DPSCs) based on highly porous nanocrystalline films of titanium dioxide [1] have attracted attention due to an alternative low-cost, high-efficiency compared to conventional silicon-based solar cells.

A typical DPSC is supported by a SnO-coated glass substrate onto which a several micrometer thick nanocrystalline metal-oxide film comprising adsorbed Ru-bipyridyl-based dye is deposited. The dye-coated metal oxide crystallites are in contact with the electrolyte solution containing iodide and triiodide ions. A second electrical contact is provided with Pt-coated SnO deposited on a glass support. The light/electrical current conversion proceeds via light absorption by the dye molecules. For an adequate energy offset between the excited levels of the dye and the minimum of the metal oxide conduction band, the excited electrons are injected from dye into the metal-oxide conduction band and collected by a transparent SnO electrode. The electrolyte reduces the oxidized dye molecules by I ions completing thereby the electric charge current.

Despite of impressive conversion efficiencies, such design suffers from a major drawback related to difficulties of encapsulation of the liquid electrolyte. Therefore, increasing effort is being invested in replacing the liquid electrolyte with a solid-state equivalent. Currently, organic semiconductors (OS) and ionic conducting polymers appear to be the most promising materials [2, 3, 4], although the efficiency of all solid state solar cells remain lower than their liquid-electrolyte-based counterparts. The principle of operation of a solid-state DPSC is based on dye excitation and electron injection into a mesoporous wide-band-gap semiconductor. The OS layer serves as a dye-regeneration medium. The excited dye molecules are regenerated by hole injection into the OS layer. The hole mobility in OS layer must be threfore relatively high, and the energy offset at the dye-OS interface favoring hole injection into the OS layer. The mesoporous structure of metal oxide plays an important role in increasing the overall conversion efficiency by increasing the area of dye/metal-oxide interface. The OS material must occupy the intergranular space of the metal oxide in order to increase the dye-regeneration probability.

This determines the method of deposition of OS layer, excluding at first sight vacuum evaporation since it yields a relatively abrupt interface between dye-sensitized metal oxide and hole transporting material. In this paper we will show, however, that vacuum evaporation of suitably chosen OS material onto a dye-sensitized nanostructured material can yield photovoltaic devices that exhibit photon-to-electron conversion efficiencies that are comparable to the solid-state DPSCs fabricated by solution-based procedures for depositing OS layers.

EXPERIMENTAL DETAILS

The solar cells were fabricated on glass substrates coated with F:SnO acting as a bottom electrode. Mesoporous SnO was dip-coated on the substrate from sols made after reacting HO with SnCl (peroxo-sol-gel-route) [5], and sensitized with cis-bis(isothiocyanato)bis(2,2'-bipyridyl-4,4'-dicarboxylato)-ruthenium(II). A 300 nm thick layer of organic semiconductor 3,4,9,10-perylene tetracarboxylic dianhydride (PTCDA) was deposited onto the sensitized mesoporous SnO by vacuum evaporation in a growth chamber with a base pressure of Torr.

During PTCDA deposition the substrate was at room temperature (RT) and a growth rate of 0.13 nm/s was employed. Commercially available PTCDA source material was purified by evaporating twice onto a collecting shutter mounted above the tungsten boat. Upon completion of PTCDA deposition, a 550 nm thick In layer was deposited on the PTCDA layer surface, and acted as a topmost electrode. Schematic diagram of the cross-sectional view of the sample is presented in figure 1.

The photoexcitation of the samples was performed by illuminating the glass substrate by a light from a 125 W Xe lamp coupled to a single-pass grating monochromator. The photocurrent was measured using a Keithley 2400 Source Meter in a short-circuit arrangement. The current-voltage characteristics were obtained with the sample placed in a dark chamber. The transient photocurrent

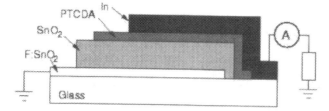

Figure 1: Schematic representation of a cross-sectional view of a solid-state dye-sensitized solar cell.

measurements were performed using a standard time-of-flight technique [6]. The samples were illuminated from the substrate side by a dye laser operated at a wavelength of 460 nm. To avoid sample damage we attenuated the 20 ns-laser pulses by using the reflected light of a glass plate.

DISCUSSION

The wavelength dependence of the incident-photon-electron-conversion efficiency (IPCE) is presented in figure 2, where we show the photocurrent measured as a function of the excitation wavelength (circles) compared to the absorption spectrum of dye-sensitized SnO (line).

The shape of the IPCE curve exhibits a doublet with components at 445nm and 475 nm, and a single peak centered at 555 nm. We emphasize that the IPCE reaches maximum values of about 1 %, which is comparable to the IPCE observed in early liquid-electrolyte-based DSPSCs [7]. Also, note that the maximum of the IPCE does not coincide with the maximum absorption coefficient of a dye-sensitized mesoporous TiO. In order to interpret the observed behavior of IPCE we consider first the structural properties of the samples.

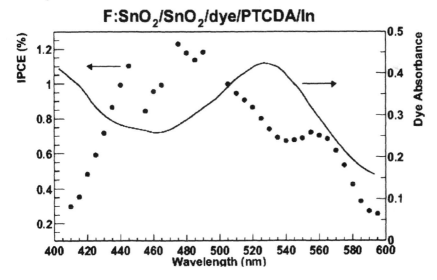

Figure 2: Incident-photon-to-collected-electron efficiency as a function of the excitation wavelength measured on a solar-cell structure comprising dye-sensitized mesoporous SnO interfaced to 300 nm thick PTCDA layer (filled circles). Absorption coefficient of a dye-sensitized metal oxide (line). The topmost and the bottommost contacts were SnO and In, respectively. The sample was illuminated from a substrate side.

Dye-sensitized mesoporous SnO layer comprises grains with size in order of 10 nm [5], and the intergrain spacing having similar dimension. Onto such surface PTCDA layers were grown. It is known [8] that the structure of PTCDA layers strongly depends on the growth conditions and the substrate structure. Disordered substrates yield higher degree of disorder in the overlayers. In addition, relatively small intergrain space precludes efficient migration of the impinging PTCDA molecules further into the mesoporous layer. We may therefore assume that the PTCDA layer exhibits a substantial degree of disorder, and that relatively thin region of the interface on the SnO side contains dye molecules contacted to PTCDA molecules. Consequently, hole injection from dye molecules into the PTCDA layer is unlikely far from the interface, even in the hypothetical case of convenient energy level .

The energy level alignment at the dye-PTCDA interface is indeed unfavorable for the hole injection into PTCDA. This is evident from figure 3, where we schematically show the electron energies of the bottom of the conduction band and the top of the valence band in SnO, the lowest unoccupied molecular orbital (LUMO) and the highest occupied molecular orbital (HOMO) in PTCDA, and the ground state and the first excited state of a single dye molecule.

The energy scale is referenced to the vacuum level. The positions of the energy levels were found in Refs. [9] and [10] for F doped SnO and dye, and PTCDA, respectively. We see that the alignment of the energy levels at the SnO/dye interface is favorable for the electron injection from the excited levels of a dye into the conduction band of mesoporous semiconductor. On the other hand the position of the highest occupied molecular orbital (HOMO) in PTCDA is about

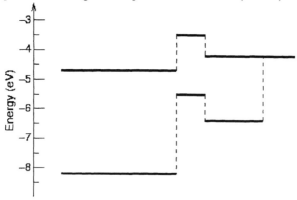

Figure 3: The electron energy levels for materials included in a dye-sensitized solid-state solar cell. The zero of the energy scale is at the vacuum level. Shown are the bottom of the conduction band and the top of the valence band in SnO, the lowest unoccupied molecular orbital and highest occupied molecular orbital in PTCDA and the ground state and the first excited state of a single dye molecule.

1.1 eV lower than the ground state of a dye. Such energy offset acts as a barrier for the hole injection from dye into PTCDA, hindering the regeneration of the excited dye molecules. On the opposite side of a PTCDA layer we have an ohmic contact to In[11] with the work function of 4.2 eV.

From the structural characteristics of our samples and electronic energy alignment at the individual interfaces it is obvious that the picture of dye regeneration via hole injection does not apply here. What is then the cause of charge separation in our PTCDA-based structures? Looking again at Fig. 3 we note that when the materials are separated in vacuum there exists a 0.5 eV difference in the work functions of SnO and In. When the materials are brought into contact the simple Schottky-Mott approach yields electrostatic potential that decreases from the F:SnO side to In side. Here we assume that a complete SnO- PTCDA region is void of intrinsic carriers. Upon illumination the mesoporous SnO acts as a reservoir of electrons excited from dye molecules which effectively decreases the position of the chemical potential on the SnO side, providing thereby a driving force for electron transport via the SnO conduction band into the PTCDA layer. The current of the electrons is therefore from the illuminated SnO side towards the In contact. This may be verified also by a dark

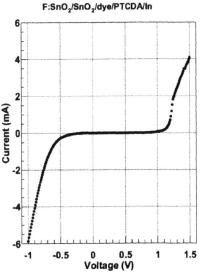

F:SnO$_2$/SnO$_2$/dye/PTCDA/In

Figure 4: Current-voltage characteristics of the PTCDA based solar-cell. The positive side of the voltage axis corresponds to the positive voltage applied to the indium (collecting) contact and negative voltage applied to the F:SnO contact.
current-voltage (I-V) characteristics of the same sample presented in Fig. 4. The positive side of the voltage axis corresponds to the positive voltage applied to the indium (collecting) contact and negative voltage applied to the F:SnO contact. We see a rectifying characteristics under the forward bias when the

electron current from SnO to In starts flowing only when sufficiently high negative bias is applied to the SnO side of the structure.

Upon illumination electrons are generated in dye-sensitized SnO. Light absorption in PTCDA is predominantly through the generation of Frenkel-type excitons[12]. The excitons drift through the PTCDA layer and eventually dissociate at the contact and contribute to the photocurrent. Qualitatively, the lineshape of the IPCE curve from Fig. 1 is similar to the absorption spectra in PTCDA as measured by Forrest and coworkers[13], exhibiting maxima in absorption at the same wavelength as we observe maxima in photocurrent. We submit that the major photogeneration occurs within the PTCDA layer, while photoexcited charge residing within the mesoporous SnO provides additional field for exciton drift and dissociation.

CONCLUSIONS

Solar-cell structures were fabricated from dye-sensitized SnO and PTCDA organic semiconductor sandwiched between transparent F:SnO and In contacts. Photon-electron conversion efficiencies of such structures are on the order of 1 %. The mechanism responsible for the operation is likely to be related to the built-in voltage due to the energy offsets between the two metallic contacts.

REFERENCES

1.   B. O'Reagan and M. Grützel. Nature, **353**, 737 (1991).
2.   U. Bach, D. Lupo, J. E. Moser, F. Weisstötel, J. Salbeck, H. Spreitzer, and Gr¨atzel. Nature, **395**, 583 (1998).
3.   K. Murakoshi, R. Kogure, Y. Wada and S. Yanagida. Chem. Lett., 471–472 (1997).
4.   K. Tennakone, G. R. R. A. Kumara, A. R. Kumarasinghe, K. G. U.Wijayantha, and P. M. Sirimanne. Semicon. Sci. Technol., **10**, 1689 (1995).
5.   U. Opara-Krašovec, B. Orel, S. Ho¡cevar and I. Muššević. J. Electrochem. Soc, **144** (1997).
6.   P. M. Borsenberger and D. S. Weiss. *Organic Photoreceptors for Imaging Systems* (Marcel Drekker, Inc., 1993).
7.   H. Lindström, H. Rensmo, S. Södergren, A. Solbrand and S.-E. Lindquist. J. Phys. Chem., **100**, 3084 (1996).
8.   S. R. Forrest. Chem. Rev., **97**, 1793 (1997).
9.   F. Akira, H. Kazujito andW. Toshiya. *TiO Photocatalysis, Fundaments and Applications* (BKC Inc., 1999).
10.   I. G. Hill, A. Rajagopal, A. Kahn and Y. Hu. Appl. Phys. Lett., **73**, 662 (1998).
11.   Y. Hirose, A. Kahn, V. Aristov, P. Soukiassian, V. Bulović and S. R. Forrest. Phys. Rev. B, **54**, 13748 (1996).
12.   Z. Shen and S. R. Forrest. Phys. Rev. B, **55**, 10578 (1997).
13.   V. Bulović, P. E. Burrows, S. R. Forrest, J. A. Cronin and M. E. Thompson. Chem. Phys., **210**, 1 (1996).

Mat. Res. Soc. Symp. Proc. Vol. 730 © 2002 Materials Research Society   V3.7

Structural and Optical Characterization of $Cu_xGa_ySe_2$ Thin Films under Excitation with Above and Below Band Gap Laser Light

C. Xue[1], D. Papadimitriou[1], Y.S. Raptis[1], T. Riedle[2], N. Esser[2], W. Richter[2],
S. Siebentritt[3], S. Nishiwaki[3], J. Albert[3], M.Ch. Lux-Steiner[3]

[1]National Technical University of Athens, Department of Physics, GR-15780 Athens, Greece.
[2]Technical University of Berlin, Institute of Solid State Physics, Hardenbergstr. 36, D-10623 Berlin, Germany.
[3]Hahn-Meitner Institute, Glienickerstr. 100, D-14109 Berlin, Germany.

ABSTRACT

$Cu_xGa_ySe_2$ MOCVD and PVD grown films were structurally and optically characterized by Raman, Micro-Raman and photoluminescence spectroscopy. Defect related photoluminescence excitation with wavelengths varying across the material band gap reveals: a) in Cu-rich $CuGaSe_2$ films, three band edge splitting due to the spin-orbit interaction and the crystal field, and donor-acceptor pair recombination between a shallow donor and two different acceptor levels, and b) in Ga-rich $CuGaSe_2$ films, donor-acceptor pair transitions between quasi-continua of donor and acceptor levels related to potential fluctuations. Raman spectra of $Cu_xGa_ySe_2$ films, excited by laser light near and below the material band gap, show intense modes at $197cm^{-1}$, $187cm^{-1}$, and $277cm^{-1}$, which can be used as indicators of crystallinity and Ga-content of the films. Polarization- and angular- dependent micro-Raman spectra of MOCVD $CuGaSe_2$ indicate that Cu_xSe_y-crystallites, dispersed on the surface of Cu-rich films, are grown oriented with their c-axis perpendicular to the film surface.

INTRODUCTION

Ternary chalcopyrite semiconductors of type Cu-III-VI$_2$ (III=Ga, In, and VI=S, Se) have attracted a lot of attention because of their potential applications in solar cell technology. $CuGaSe_2$, as a higher band-gap chalcopyrite (E_g=1.68 eV at 300K), can be used, in combination with $CuInSe_2$ (Eg≈1eV), in the growth of alloyed materials[1] with band-gap energies over a wide spectral range. For good cell performance, the properties of its constituting materials have to be well understood. In this work, the structural and optical properties of $CuGaSe_2$-films grown on Glass/Mo-substrates by commercially attractive PVD techniques were studied with reference to the properties of $CuGaSe_2$ epitaxial layers grown by MOCVD on GaAs (100).

EXPERIMENT

All the studied samples were grown at Hahn-Meitner-Institute in Berlin as described elsewhere[2-3]. Photoluminescence (PL) emission was excited in the temperature range 300-20 K by 514.5 nm(2.4 eV) Ar$^+$-laser and Ar$^+$-laser pumped Ti:Sapphire-laser operated in the spectral range 1.51-1.77 eV. Raman spectra were excited by laser light near (647.1 and 676.4 nm Kr$^+$-laser) and below the material band gap (Ar$^+$-pumped Ti-Sapphire laser). The emitted light was spectrally analyzed by a double grating monochromator (SPEX 1403) with cooled PMT-detector. In addition, the Cu_xSe_y-phase formed on the surface of Cu-rich films was studied by micro-

Raman spectroscopy under excitation with 647.1 nm Kr$^+$-laser using a Jobin-Yvon T64000 triple monochromator equipped with an Olympus microscope and liquid-N$_2$ cooled CCD detector.

RESULTS AND DISCUSSION

Raman- and PL-spectra were obtained at 20 K on PVD gradient composition Cu$_x$Ga$_y$Se$_2$ (Cu/Ga=1.45-075). The film composition was estimated by energy dispersive X-ray spectroscopy (EDX). As shown in Figs.1a and 1b, in near stoichiometric CuGaSe$_2$ modifications, significant changes occur in both Raman- and PL-spectra by changing from the Cu-rich to the Ga-rich phase. PL-data are consistent with previously published data of these films[3].

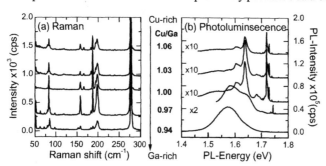

Figure 1. Raman and PL spectra of PVD gradient composition CuGaSe$_2$ films at 20 K excited by 676.4 nm Kr$^+$-laser.

Band schema

The PL-spectra of MOCVD Cu-rich samples excited by 514.5nm Ar$^+$-laser include contributions of three components emitted in the range 1.4-2.1 eV as demonstrated in Fig. 2: 1) Spectra originated via valence band splitting have three edge emission bands at 2.02, 1.81 and 1.71 eV, 2) Spectra due to donor-acceptor-pair transitions consist of two strong emission peaks at 1.66 and 1.63 eV and a weak peak at 1.60 eV identified as phonon-replica[4], 3) PL-emission from the GaAs-substrate appears at 1.49 and 1.51 eV.

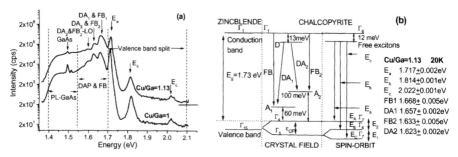

Figure 2. Cu-rich Cu$_x$Ga$_y$Se$_2$: (a) PL-spectra of MOCVD samples at 20 K (in log. scale), and (b) band schema (acceptor levels determined from FB transitions, donor level as discussed in text).

With increasing Ga content, band-to-band and donor-acceptor-pair transitions (DAP, 1.63 and 1.66 eV) between discrete donor and acceptor levels diminish, and quasi donor-acceptor-pair recombination (QDAP centered at 1.62-1.55 eV depending on composition) takes over.

For a more detailed study of PL-emission and precise identification of PL-bands and their origin, a variety of excitation energies were used (Fig. 3a). In particular, by excitation with a laser beam energy below the material band gap (1.73 eV at 20K), contributions of DAP-transitions to PL-emission could be revealed. By decreasing the excitation energy from 1.77 to 1.57 eV, DA and FB emission peaks decrease. As demonstrated in Figs. 3a and b, 1) PL-bands at 1.66 and 1.63 eV are composed of two peaks assigned[3] to free-to-bound transitions (high energy side) and to donor-acceptor-pair recombination (low energy side), with $E_{FBi}-E_{DAi} = (10\pm2)$ meV, i=1,2, consistently for both bands, 2) with decreasing excitation energy, the intensity ratio of DA to FB increases (Fig. 3b for DA1), and the band characteristics of the DA-transition are revealed (Fig. 3a), 3) the intensity of the DA1 transition as a function of the excitation energy (Fig. 3b) is maximum at excitation energy 1.717 eV, that is 13 meV below the conduction band edge, while even small excitation energy changes cause a rapid decrease of the DA1 emission. The maximum emission has, therefore, to be considered as a result of resonance effects. At excitation energy 1.703 eV the intensity of the DA1 emission decreases considerably due to the inability of the low energy pumping beam to excite electrons from valence band to donor state. The DAP2 transition exhibits energy and intensity changes in dependence of excitation energy similar to those of the DAP1.

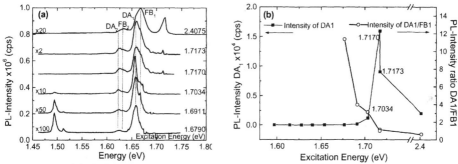

Figure 3. Cu-rich CuGaSe₂ samples (Cu/Ga=1.13): (a) PL spectra for excitation energies 2.41-1.68eV, and (b) dependence of DA1/FB1 intensity ratio, and DA1-intensity on excitation energy.

The observed dependence of the DAP1 transition on excitation energy justifies positioning of the donor level, in Fig. 2b, at a distance of 13 meV below the lower conduction band. Moreover, the fact that evaluation of FB- and DAP-transitions yields an energy difference of $\Delta E_{FB-DAP} = (10\pm2)$ meV is in perfect accordance with this donor depth. This result is the exact determination of the energy of the shallow donor proposed by Bauknecht[3].

PL spectra of Ga-rich MOCVD samples were also recorded at 20 K by excitation with different wavelengths of the Ti:sapphire laser. The emitted PL exhibits a strong red energy-shift and intensity decrease for excitation energies between 1.68eV and 1.54eV, while the band width is almost constant (Figs. 4a and 4b). This behavior can be explained by the model of fluctuating potentials due to the high degree of compensation in Ga-rich samples[3]. By lateral integration one obtains a quasi-continuum of donor and acceptor states apparently shifted closer in energy by

50meV. This corresponds to an amplitude of the potential fluctuations of 50meV which occurs for a compensating donor concentration of 2.5×10^{18} cm^{-3}. For excitation energies around 1.64 eV the donor states are pumped directly and relax to the underlying acceptor states. For excitation energies less than 1.64 eV, the only possibility to populate these states is by exciting carriers from the quasi-continuum of acceptor states that can be still reached by the laser beam. Donor states, pumped this way, may then relax back to the ionized acceptor states. Further decrease of the excitation energy leads to the excitation of the less populated upper acceptor states, which results in a decrease of both the energy and the intensity of the corresponding PL-bands. The band-width remains almost constant due to the quasi-continuous nature of the contributing energy states.

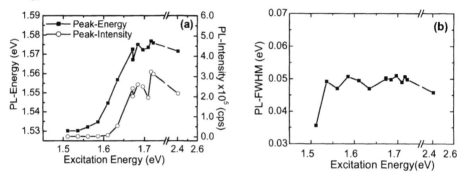

Figure 4. Ga-rich CuGaSe$_2$ samples (Cu/Ga=0.96): Dependence of (a) PL-band energy and intensity, and b) PL-band width on excitation energy.

Structural characterization

The temperature dependence of the first-order Raman scattering offers an important tool to check specific anharmonic properties. The temperature dependent frequency changes, $\partial\omega/\partial T$, of the E1- and A1-modes of MOCVD samples is larger in the Ga-rich modifications of CuGaSe$_2$, which supports the presence of anharmonic effects in the Ga-rich films. By comparison of the temperature dependent frequency shifts of the Raman modes of Ga-rich films at 187(A$_1$), 197(B$_2$), and 277(E$_1$) cm^{-1}, it is evident that: $\partial\omega/\partial T$(B2) > $\partial\omega/\partial T$(E1) > $\partial\omega/\partial T$(A1). The strong anharmonic character of the B2-mode, further supported by the temperature dependent changes of the band width of this mode, can be used as indicator of defect related anharmonicity effects.

In Cu-rich samples, independently of composition, the emission energy of PL bands and the frequency of Raman modes do not significantly change. This is due to the fact that Cu-rich modifications are composed of almost stoichiometric CuGaSe$_2$. The Cu-excess forms a secondary Cu$_x$Se$_y$-phase on the film surface, and does not increase the Cu-content of the chalcopyrite phase. In near stoichiometric CuGaSe$_2$ modifications, the intensity of the B$_2$ mode at 197 cm^{-1} increases and exceeds that of the A1 mode at 186 cm^{-1}. The intensity-ratio of the B2- to A1-mode, is, in general, I(197cm^{-1})/I(187cm^{-1}) < 1 in Cu-rich and I(197cm^{-1})/I(187cm^{-1}) > 1 in Ga-rich modifications, except of the Cu-highly-rich modifications where it is affected by the presence of the Cu$_x$Se$_y$-phase. Changes in phonon intensities appear simultaneously with PL-band shape and intensity changes (Fig. 1), and are related to modifications of the electronic properties of the samples caused by the stoichiometry change. Since chalkopyrite is a low symmetry crystal the phonon modes of different symmetry may couple to different electronic bands. Stoichiometry dependent defect generation may influence these electronic bands in a different way, thus leading to the observed intensity changes. Moreover, in Ga-rich samples, with the increase of Ga content, the A1- and E1-Raman modes, at 187 and 277 cm^{-1} respectively, are broadened. The E1-mode is the most intensive one and its band width can be exactly estimated. Band broadening with increasing Ga-content is indicative of increasing disorder and is correlated to an increasing number of defects and a higher efficiency of defect related PL-bands.

According to the above, the E1-mode can act as calibrator of the Ga-content of CuGaSe$_2$-films, while the intensity-ratio of B2- to A1-mode can be used as indicator of crystal quality.

Surface structural characterization

In the micro-Raman spectra of the studied films, recorded by focusing the laser beam on a Cu$_x$Se$_y$ crystallite, new Raman modes appeared at 13, 45, and 263 cm^{-1} in addition to the Raman modes of CuGaSe$_2$ (Figs. 5a and 5b). These modes are in good agreement with the Raman modes observed[5] on crystalline CuSe, which belongs to the hexagonal space group D_{6h}^4 and has eight active Raman modes: $2A_{1g}+2E_{1g}+4E_{2g}$. However, there may exist several structural phases of Cu$_x$Se$_y$ in the studied films. In XRD, Cu$_9$Se$_5$ is detected, but this has a rather small Raman scattering cross-section.

Cu$_x$Se$_y$-crystallites dispersed within an area of 50x50 μm were measured, under parallel incident-scattered polarization and variable sample orientation, taking as a reference for the angle-measurements the (110) cleaved edge of the GaAs-substrate. Since the intensity variations can be due to both the angular dependence of Raman scattering on crystallite surface and the morphology of crystallite surface (with typical size in the order of the laser spot), in order to estimate the dependence of Raman intensities on the angle of rotation, the intensities of the 45cm^{-1}–mode of Cu$_x$Se$_y$ and the 82cm^{-1}–mode of CuGaSe$_2$, were normalized with respect to the A$_{1g}$-mode (263 cm^{-1}) of Cu$_x$Se$_y$ and the A$_1$-mode (185 cm^{-1}) of CuGaSe$_2$ respectively, for which no angular dependence is expected. The angular dependence of the normalized intensities I_{45}/I_{263} and I_{82}/I_{185} is shown in Figs. 5c and 5d, respectively. In Fig. 5c, the intensity ratio as a function of angle varies around a mean value indicating that the 45cm^{-1}–mode is of the same symmetry as the 263cm^{-1}(A$_{1g}$)–mode. On the other hand, in Fig. 5d, there is a systematic angular variation as expected for a B$_2$ mode, which can be fitted with an expression of the type: $C+Asin^2(2\theta-2\theta_0)$, if it is assumed that the crystallographic axes of the tetragonal system are rotated by an angle θ_0 with respect to the cleaved face. The result of the fitting, shown in Fig. 5d as a solid line with a fitting parameter θ_0=42,0°±3,2°, confirms both, the B$_2$ character of the 82cm^{-1}–mode, and the coherent growth of the CuGaSe$_2$ epilayer with the crystallographic axis (100) parallel to (100) axis of the GaAs substrate.

Measurements performed on several Cu$_x$Se$_y$ microcrystallites in crossed polarization configuration (with polarizer and analyzer oriented parallel to the (100)- and (010)-axis of the substrate respectively) result in significant decrease (or even complete elimination) of scattering intensities related to Cu$_x$Se$_y$-modes. This is a strong indication of both the A$_{1g}$ character of these modes and the preferential orientation of the crystallites, in agreement with the SEM images of the films. The preferential growth of Cu$_x$Se$_y$-crystallites on CuGaSe$_2$ surface, with the c-axis of the crystallite hexagonal symmetry perpendicular to the epilayer, is further confirmed by the strict application of the selection rules for the B$_2$ mode of the CuGaSe$_2$-epilayer at 82 cm^{-1} (Figs.

5a and 5b), though the incident and the scattered light are passing through crystallites with a typical thickness in the order of magnitude of the excitation wavelength. For any other orientation of Cu_xSe_y, the incident and the scattered light would have suffered polarization scrabbling resulting in an apparent violation of the polarization selection rules.

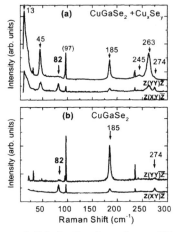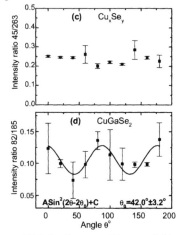

Figure 5. Polarization dependence of Raman spectra of (a) Cu_xSe_y crystallites, and (b) $CuGaSe_2$-film. Angular dependence of mean intensity ratio of (c) $CuGaSe_2$-modes, and (d) Cu_xSe_y-modes.

CONCLUSIONS

In the present study, the band schema suggested[3] for defect related optical transitions in $CuGaSe_2$-films was experimentally confirmed by PL excitation across the material band gap. The energy of the shallow donor could be specified. The film structure was characterized according to Raman spectra excited by laser light near and below the material band gap. Raman modes at $197 cm^{-1}$, $187 cm^{-1}$, and $277 cm^{-1}$, can be used as indicators of defect related anharmonicity effects, crystallinity and Ga-content of the films respectively. Polarization- and angular- dependent micro-Raman measurements of MOCVD $CuGaSe_2$-films showed that the $CuGaSe_2$-film and the Cu_xSe_y crystallites on film surface are grown on the GaAs-substrate with a preferential orientation.

REFERENCES

1. M. Contreras, B. Egaas, K. Ramanathan, J. Hiltner, A. Swartzlander, F. Hasoon, and R. Noufi, *Prog. Photovoltaics* **7**, 311 (1999).
2. S. Schuler, S. Nishiwaki, M. Dziedzina, R. Klenk, S. Siebentritt, M. Ch. Lux-Steiner, MRS Proc. **668** (2002) H5.14.
3. A. Bauknecht, S. Siebentritt, J. Albert, and M. Ch. Lux-Steiner, *J. Appl. Phys.* **89(8)**, 4391 (2001).
4. A. Bauknecht, S. Siebentritt, J. Albert, Y. Tomm, and M. Ch. Lux-Steiner, *Jap. J. Appl. Phys.* **39**, suppl. 39-1, 322 (2000).
5. M. Ishii, K. Shibata, H. Nozaki *J. Solid State Chem.* **105**, 504 (1993).

Mat. Res. Soc. Symp. Proc. Vol. 730 © 2002 Materials Research Society   V3.8

Direct-Write Printing of Silver Metallizations on Silicon Solar Cells

C. J. Curtis, T. Rivkin, A. Miedaner, J. Alleman, J. Perkins, L. Smith, and D. S. Ginley
National Renewable Energy Laboratory, Golden, CO 80401

ABSTRACT

Direct-write technologies offer the potential for low-cost materials-efficient deposition of contact metallizations for photovoltaics. We report on the inkjet printing of metal organic decomposition (MOD) inks with and without nanoparticle additions. Near-bulk conductivity of printed and sprayed metal films has been achieved for Ag and Ag nanocomposites. Good adhesion and ohmic contacts with a measured contact resistance of 400 $\mu\Omega\bullet cm^2$ have been observed between the sprayed silver films and a heavily doped n-type layer of Si. Silver deposited using the MOD ink burns through the Si_3N_4 antireflection coating when annealed at 850°C to form an ohmic contact to the n-Si underneath. An active solar cell device was fabricated using a top contact that was spray printed using the Ag MOD ink. Inkjet printed films show adhesion differences as a function of the process temperature and solvent. Silver lines with good adhesion and conductivity have been printed on glass with 100 μm resolution.

INTRODUCTION

A key area for improvement of photovoltaic cells is the development of low-cost materials-efficient process methodologies. Atmospheric process approaches potentially offer these advantages. Inkjet printing, as a derivative of direct-write processing, offers the additional advantages of low capitalization, very high materials efficiency, elimination of photolithography, and noncontact processing [1].

Conceptually, for Si solar cells, all device elements except the Si could be directly written or sprayed, including contact metallizations (front and rear), dopants, transparent conductors, and antireflection coatings. Our initial thrust has been in the area of developing contact metallizations. As the thickness of Si cells falls below 100 μm, contact grids for the front and rear contacts can be inkjet printed, even on the rough surface of polysilicon, without contacting the thin, fragile substrates. At present, inkjets are capable of line resolutions < 20 μm, which is at least two times better than the current state of the art obtained by screen printing [2]. In addition, it is an inexpensive, atmospheric process and can be an environmentally friendly, no-waste approach.

A major challenge in applying inkjet processes for direct writing is formulating suitable inks. The inks must contain the appropriate precursors and a carrier vehicle. In addition, they may contain various binders, dispersants, and adhesion promoters, depending on the nature of the precursor and the particular application. In the case of inks for metallization, the content of the metallic ink must be adjusted to provide the required resolution, with good adhesion and the desired electronic properties for the conducting lines. Ink composition is critical because it defines the way in which the ink can be jetted, the adhesion to the substrate, and the line resolution and profile, and it can control the mechanism of metal formation.

Our specific goals are to develop inks and optimize printing parameters for highly conductive lines, achieving low contact resistance, good adhesion, and high resolution.

Approach

Overall, we have chosen to use combinations of organometallic metal precursors with metal nanoparticles. This allows essentially a mix-and-match approach between the MOD precursor and various nanoparticles which can be tailored for a particular application, e.g. doping level, thickness, and process temperature. We use organometallic precursors as both a metal-forming component of the ink and as a "glue" to bond the nanoparticles together and to enhance adhesion to the substrate. The dry organometallic compound (such as silver(hexafluoroacetylacetonate)-(1,5-cyclooctadiene) or Ag(hfa)(COD)) is dissolved in an organic solvent such as toluene, ethanol, or butanol. The ink is delivered by spraying or is inkjet printed on a heated substrate in the desired pattern. A silver film forms upon solvent evaporation and decomposition of the printed precursor at elevated temperature (~300°C). Gaseous byproducts of decomposition leave the system, providing contamination-free metal films. To increase the silver loading of the ink and obtain higher deposition rates, silver or other metal nanoparticles may be added to the ink along with the organometallic precursor. In this configuration, silver particles comprise the main conducting volume of the resultant coating, while the organometallic constituent acts as a glue for the silver particles, providing enhanced electrical and mechanical bonding of the metal particles with the substrate and between themselves. Fine, deagglomerated nanoparticulate metal powders must be used in this ink so as to avoid clogging the 10–50 μm orifice of the inkjet. In addition, active constituents, such as adhesion promoters, surface activators, precursors of n-dopants for selective emitters, or possibly nanosized glass frits, may be added to the ink to achieve the required electronic and mechanical properties of the contact.

For initial evaluation of new ink compositions, we have used spray deposition. In this technique, droplets of ink are deposited on heated substrates using an airbrush. This deposition technique is very similar to inkjet printing, but simpler and more robust. The relatively large nozzle opening of the airbrush makes it more difficult to clog. The main difference from the inkjet is that spray deposition is best suited for deposition of continuous films over large areas and is not spatially selective. Practice has shown that inks that spray well by this approach tend to work well in the inkjet as well.

EXPERIMENTS and RESULTS

Spray Deposition with Ag Metal-Organic Ink. A saturated solution of Ag(hfa)COD in toluene, filtered through a 0.2 μm syringe filter, was used as the metal-organic Ag ink [3]. The ink was sprayed onto heated substrates in air using a hand-held Vega 2000 airbrush. During depositions, the substrates were attached using metal clips to a heated copper plate maintained at 400°C. Smooth, dense, grainy coatings were obtained on glass and silicon substrates, as can be seen in the scanning electron microscope (SEM) images in Figure 1. The thickness of the coatings deposited ranged from 0.1 to 4 μm. According to X-ray photoelectron spectroscopy (XPS) the chemical content of the coatings was pure silver. A significant level of oxygen, fluorine, and carbon impurities was detected at the film surface and is most likely due to post-deposition surface contamination. It rapidly decreases with depth during ion milling, and the film is >99 atom-% Ag after removing 300 Å. X-ray diffraction (XRD) analysis confirmed that the coating as-deposited consisted primarily of (111)-oriented silver grains, with a small fraction having (100) and (110) orientation. The adhesion strength of the sprayed 3.4 μm thick silver coating on Si exceeded the adhesion strength of conventional screen-printed contacts. Resistivity of the sprayed silver layer (2 μΩ•cm) was very near that of bulk silver. Significantly,

 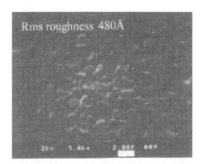

Figure 1. SEM images of Ag films spray printed on glass (left) and Si (right) using ink prepared from Ag(hfa)(COD).

we were able to achieve ohmic contact with a relatively low contact resistance (400 $\mu\Omega\bullet cm^2$) [4] between the sprayed silver coating and the n-type layer of a diffused p-n junction silicon solar cell. The contact resistance was measured using the Transfer-Length Method (TLM) [5]. The silver pads for TLM were fabricated from the sprayed coatings using photolithography, followed by an etch with 50 % nitric acid.

These results are significant because films with high adhesion strength and good electronic properties were achieved with a simple, one-step, low-temperature deposition process. This is in contrast to screen printed coatings, which require addition of glass frits to promote adhesion and high annealing temperatures (~700°C), or vacuum deposited coatings, which are expensive and require an additional adhesion layer, to achieve similar properties.

Spray Deposition with Composite Ag Nanoparticle/Metal-Organic Ink. Composite nanoparticle/metal-organic ink was prepared by mixing 1.0 g Ag nano-particles in a solution formed by dissolution of 1.0 g Ag(hfa)(COD) in 4 ml ethanol (1:3.9 molar ratio). This precursor was spray printed on a glass substrate heated to 300°C. As deposited, the 10 μm thick silver layer had a relatively porous grainy structure (Figure 2). The layer in Figure 2 was characterized by good adhesion to a Pyrex microscope slide, which was confirmed with the Scotch tape pull test (15-20 lb/in). The high initial resistivity of the as-deposited silver layer (58 $\mu\Omega$-cm) dropped to 2.4 $\mu\Omega$cm, near the value for bulk silver, after annealing for 30 minutes in air at 400°C. The high deposition rate, along with the excellent conducting properties of the deposited silver layer from the composite ink are encouraging and we are planning to proceed to the next step – to evaluate the "printability" of the composite ink by inkjet. The primary concern is that agglomerates of the Ag particles may clog the small 50 μm orifice of the current inkjet. Also, we need to optimize the ink composition to achieve stable particle suspension. When the Ag particles are simply mixed into the organometallic ink, they tend to sediment on the walls of the container. This may require a capping agent for the Ag nanoparticles, which would potentially solve both the agglomeration and printing problems at the same time.

Direct-Write Ag by Inkjet Printing
Epson Printer. Simple test patterns were printed on glass and Si using the Ag(hfa)(COD) metal-organic ink with an Epson Stylus Color 740 piezoelectric inkjet printer modified to print labels on compact discs. The ink was prepared according to the recipe above, with ethanol replacing toluene as the solvent. Toluene was found incompatible with the plastic of the inkjet

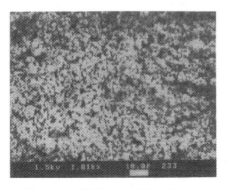

Figure 3. Single line pattern printed on glass with the Epson ink jet printer using Ag(hfa)(COD)-based ink. The width of the line is 200 μm.

Figure 2. SEM image of Ag film spray printed on glass at 300°C using the composite ink.

cartridges. The substrate was supported on an aluminum disk heated to 100°C to facilitate solvent evaporation and to prevent the ink from running on the surface. Three layers were printed on top of each other to build a reasonably thick silver coating. At the end of the deposition, the sample was annealed in air at 300°C for 20 minutes. This final annealing eliminated dark brown spots that could be seen in the coating just after the deposition. We assume that these were Ag(hfa)(COD) spots that decomposed, forming Ag during the annealing step. The average thickness of the trilayer coating was ~ 2000 Å, giving a deposition rate of ~700 Å per layer. Minimum feature size observed in the test pattern was 200 μm. The surface morphology of the inkjet printed silver observed by atomic force microscopy (AFM) closely resembles the silver layers deposited by spraying, supporting the analogy between spray and inkjet printing. The resistivity of the coating measured by the 4-probe technique was ~5 μΩ•cm, which is a little over a factor of two higher than the resistivity of bulk silver. Good adhesion of the silver coating to the glass slide was confirmed with the Scotch tape pull test.

The disadvantage of using the commercial Epson printer is in lack of control over the printing conditions used. The Epson is optimized for water-based color inks. Using the same printing conditions with the ethanol-based ink resulted in reduced line quality. An optical micrograph of a single line printed on a glass slide is shown in Figure 3. It was observed that the silver accumulated at the edges of the line. This phenomenon may be understood by considering the dynamics of liquid perturbation during the processes of drop evaporation [6].

Microfab Inkjet System. The Microfab inkjet printing system is essentially a drop-on-demand piezoelectric inkjet where user control of waveform, frequency, amplitude, line pressure, and orifice size are all possible. Drops form by voltage pulses applied to a piezoelectric actuator that creates an acoustic wave in the body of a glass capillary filled with ink. The stimulated waves break the ink meniscus off the tip of the jet at the frequency of the pulse generation. By controlling the parameters of the voltage pulse and printing variables such as the frequency of drop generation, the volume/size of the drops and the speed of the drop at ejection can be controlled. Other advantages of the new system are that the jet system does not have plastic parts and thus gives us more freedom for choosing solvents and printing at higher temperatures. Heads are available for high-temperature and nitrogen-purged printing. For the initial printing

Figure 4. Single line pattern printed on glass with the Microfab ink jet printer using Ag(hfa)(COD)-based ink. The width of the line is 200 μm.

Figure 5. Individual 100 μm Ag drops printed on glass at 200°C with the Microfab ink jet printer using Ag(hfa)(COD)/butanol ink.

experiments, however, we used the same ethanol-based ink as employed in the Epson experiments above. Because of the relatively low boiling point of the ethanol, we had to choose a high pulse frequency (1500 Hz) and a relatively low substrate temperature (150°C) to maintain an uninterrupted jet of drops. Single-line patterns were written on a heated glass substrate by linear motion of the inkjet print-head attached to a computer-controlled X-Y table. The jet-to-substrate distance was maintained at ~1 cm. To convert printed precursor lines to metallic silver, the printed substrates were heated with a heat gun to above the decomposition temperature of the organometallic compound. The lines obtained had better silver film uniformity and the same (200 μm) line width as those printed with the conventional Epson printer. Figure 4 illustrates a line printed with the Microfab system. The Ag is uniform across the line width. The orifice size in the Microfab inkjet system is 2.5 times larger than that of the Epson printer (50 μm vs. 20 μm), and yet, the widths of the lines obtained were the same. The larger orifice size, which also delivers larger drops, gave a more uniform line with better resolution (relative to orifice size) because the Microfab system can tolerate higher substrate temperatures. At higher substrate temperatures, solvent evaporation occurs more rapidly and the ink does not have time to spread as much, leading to higher resolution. These results indicate that the Microfab inkjet should be capable of much better resolution if a smaller orifice is used. (Note that heads are available from Microfab with orifices down to 10 μm.)

In our most recent printing experiments, we were able to achieve 100 μm wide lines by switching to butanol-based inks. Replacing ethanol with butanol, a solvent with higher boiling temperature, allowed a reduced drop ejection rate, higher substrate temperature, and shorter jet-to-substrate distance to be used. The combination of these changes resulted in better resolution by a factor of two. An additional benefit of using the butanol-based ink is that we were able to print on a substrate heated to well above the decomposition point of the organometallic compound, thus eliminating the need for a post-deposition annealing step. Figure 5 shows individual 100 μm drops of Ag deposited at 200°C on glass with the Microfab system using Ag(hfa)(COD) dissolved in butanol.

Spray Printed Si Solar Cell

Silicon substrates with a diffused p-n junction and coated with a Si_3N_4 antireflection (AR) layer were provided by Evergreen Solar. A silver contact grid was then deposited using the Ag(hfa)(COD)/toluene ink in the following manner. A continuous layer of Ag approximately 4 μm thick was spray deposited onto the substrate at 400°C, as described above. The contact grid

Figure 6. Optical micrograph of Ag printed on Si_3N_4-coated Si, annealed at 850°C and then etched off showing that Ag burns through the AR coating.

pattern was then developed using a photoresist process and the unwanted Ag was etched away with HNO_3. The Al back contact was then deposited on the back side by vacuum evaporation and the structure was annealed in air at 850°C. During this heating step, the Ag burns through the Si_3N_4 AR coating and forms an ohmic contact to the n-Si below. Figure 6 shows an optical micrograph of a Si substrate treated in this manner and then etched with HNO_3 to remove the Ag. The area covered by the contact is quite distinct from the uncovered Si_3N_4 area. The contact resistance of contacts formed in this way was 4.1 mΩ•cm^2. Unfortunately, after firing at 850°C the Ag grid layer was no longer continuous, but had migrated to form islands of Ag. To form a continuous contact, another layer of Ag was spray printed on top of the fired cell at 400°C and patterned using photoresist and etching, as before. This produced an active Si device with an efficiency of 6.8%. This crude experiment demonstrates the feasibility of printed Ag contacts on Si solar cells and shows that the metal-organic ink is capable of burning through the AR coating to form a low-resistance ohmic contact.

CONCLUSION

We have demonstrated that direct MOD or composite nanoparticle/MOD deposition of Ag on glass and Si produces films with conductivities comparable to vacuum-deposited materials. Inkjet printing of the precursors produced line resolutions comparable to those obtained by current screen printing technology by a non-contact approach. The deposition rates can be very high, especially for the composite inks. Silver deposited using the Ag(hfa)(COD) ink burns through the Si_3N_4 AR coating on Si substrates at 850°C and forms an ohmic contact to the Si layer underneath. An active Si solar cell was fabricated using a spray printed Ag contact grid.

REFERENCES

[1] K.F. Teng and R.W. Vest, *IEEE Electron Device Lett.,* **9**, 591 (1988).
[2] M. Ghannam, S. Sivoththaman, J. Poortmans, J. Szlufcik, J. Nijs, R. Mertens, and R. Van Overstraeten, *Solar Energy,* **59**, 1-3, 101-110 (1997).
[3] C.J. Curtis, A. Miedaner, T. Rivkin, J. Alleman, D.L. Schulz, and D.S. Ginley, *Mater. Res. Soc. Symp. Proc.,* **624**, (2001).
[4] J. Nijs, E. Demesmaeker, J. Szlufick, J. Poortmans, L. Frisson, K. De Clercq, M. Ghannam, R. Mertens, and R Van Overstraeten, *Solar Energy Materials and Solar Cells,* **41/42**, 101-117 (1996).
[5] H.H. Berger, *J. Electrochem. Soc.,* **119**, 4, 507 (1972).
[6] S. Maenosono, C. D. Dushkin, S. Saita, and Y. Yamaguchi, *Langmuir,* **15**, 957-965 (1999).

Mat. Res. Soc. Symp. Proc. Vol. 730 © 2002 Materials Research Society   V3.9

Textured zinc oxide by expanding thermal plasma CVD: the effect of aluminum doping

R. Groenen, E.R. Kieft, J.L. Linden[1], M.C.M. van de Sanden
Eindhoven University of Technology, Dept. of Applied Physics, P.O. Box 513, NL-5600 MB
Eindhoven, The Netherlands
[1]TNO TPD, Division Materials Research and Technology, P.O. Box 595, NL-5600 AN
Eindhoven, The Netherlands

ABSTRACT

Aluminum doped ZnO films are deposited on glass substrates at a temperature of 200°C by
expanding thermal plasma CVD. Surface texture, morphology and crystal structure have been
studied by AFM, SEM and XRD. A rough surface texture, which is essential for application as
front electrode in thin film solar cells, is obtained during deposition. The addition of aluminum
as a dopant results in distinct differences in film morphology, a transition from large, rounded
crystallites to a more pyramid-like structure is observed. The structure of films is hexagonal with
a preferred crystal orientation in the faces (002) and (004), indicating that films are oriented with
their c-axes perpendicular to the substrate plane. In addition, spectroscopic ellipsometry is used
to evaluate optical and electronic film properties. The presence of aluminum donors in doped
films is confirmed by a shift in the ZnO band gap energy from 3.32 to 3.65 eV. In combination
with reflection and transmission measurements in the visible and NIR ranges, film resistivities
have been obtained from the free-carrier absorption. These results are consistent with direct
measurements. Resistivities as low as $6.0 \cdot 10^{-4}$ Ωcm have been obtained.

INTRODUCTION

Zinc oxide (ZnO) is a transparent conducting oxide (TCO) of considerable technological interest
for amongst others application in thin film solar cells. It allows for superior transparency, lower
deposition temperature, lower costs and less environmental impact compared to the widely used
fluorine doped tin oxide (SnO_2:F) which shows an excellent surface texture enabling effective
light trapping properties [1]. Recently, a new method has been presented for low temperature
deposition of surface textured ZnO utilizing an expanding thermal argon plasma created by a
cascaded arc [2]. It has been shown that high quality material is obtained, showing excellent
performance in thin film a-Si:H solar cells [3,4]. Here, the effect of aluminum doping on the
ZnO film properties is investigated. The structural properties will be discussed briefly. Both
optical and electronic film properties are studied using spectroscopic ellipsometry in
combination with reflection and transmission measurements in the visible and NIR ranges.

EXPERIMENTAL

Undoped and aluminum doped ZnO films are deposited on Corning 1737F glass substrates (100x50 mm^2) by the expanding thermal plasma chemical vapor deposition process as described in detail elsewhere [2]. (Co)precursors are oxygen, diethylzinc (DEZ), and additionally for doped films, trimethylaluminum (TMA). To study the effect of aluminum doping, two series of films were deposited at 200°C, both with a variable aluminum precursor dosage in the range of 0 – 7 at.% of the total metalorganic precursor dosing. The first series with film thicknesses around 1000 nm served to study the effect of Al doping on both structural and electrical film properties. Because of the relatively large thickness and especially large surface roughness scale, it is not straightforward to determine optical constants for these films accurately with spectroscopic ellipsometry. Therefore, a second series with film thicknesses around 100 nm was deposited to determine the optical constants. It is assumed here that the aluminum percentage incorporated in thin films equals that in the thick films.

Surface texture, morphology and crystal structure were studied using atomic force microscopy (AFM), scanning electron microscopy (SEM) and X-ray diffraction (XRD). Film thickness and sheet resistance were determined with a step profiler and a van der Pauw four-point probe, respectively.

Spectroscopic ellipsometry measures the change in polarization of light as a function of wavelength when light is reflected from a sample. This polarization change is expressed in terms of the ellipsometric parameters Ψ and Δ which are related to the total Fresnel reflection coefficients R_p and R_s respectively parallel and perpendicular to the plane of incidence including interference effects [5]. Ellipsometric Ψ and Δ data were acquired at an angle of incidence of 75° over the spectral range of 245 – 1000 nm with a resolution of 1.6 nm using a Woollam M-2000U rotating compensator ellipsometer.

To determine the optical constants in the infrared part of the spectrum, ellipsometric measurements were combined with reflection and transmission measurements, using a Perkin Elmer Lambda 900 UV/VIS/NIR Spectrometer with a Pela 1020 60 mm integrating sphere. Transmission was measured in a straight-through configuration, whereas reflection was measured under an angle of incidence of 8° with respect to normal. The wavelength range of these measurements is 250 – 2500 nm.

RESULTS AND DISCUSSION

In contrast to earlier results [3], *stoichiometric* ZnO is deposited as the undoped reference material. A decreasing deposition rate from 1 nm/s to 0.6 nm/s is observed with increasing aluminum precursor dosage. SEM micrographs indicate that both undoped and aluminum doped films are strongly textured, growing in a columnar structure. Like deposition temperature and plasma power, the presence of aluminum as a dopant has a peculiar influence on film morphology, as shown in Fig. 1. A transition from large, rounded crystallites to a more pyramid-like structure is observed. The surface texture of the doped films also exhibits some vertical steps and sharp peaks, which might lead to the observed shunting problems in a-Si:H solar cells [6]. AFM measurements are in agreement with these observations, revealing root mean square roughnesses of 40 to 45 nm. XRD diffractograms confirm the hexagonal wurtzite structure. A preferred observation of (002) and (004) peaks indicate films are oriented with their c-axis perpendicular to the substrate plane. At 7 at.% aluminum precursor dosage, additional orientations in the faces (101) and (100) appear. No alumina phase is detected in the XRD patterns, which suggests aluminum replaces zinc substitutionally in the hexagonal lattice or aluminum segregates to the non-crystalline region in grain boundaries.

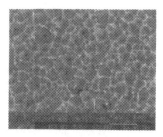 

Figure 1. SEM images of the surface of thick undoped ZnO (left) and aluminum doped ZnO with 2 at.% aluminum precursor dosage (right).

A simple but appropriate optical model for the visible wavelength range commonly applied for ZnO is the Sellmeier model, consisting of two ideal harmonic oscillators with different resonance frequencies, one in the UV for the above-band edge absorption and one in the IR for the free carrier absorption [7]. To fit the combined ellipsometric and reflection and transmission data, the last term is replaced by a Drude oscillator, which is a single Lorentz oscillator with its center energy fixed at zero [8].

The optical constants of the bare glass substrate are determined seperately. The model is used to determine film thickness and roughness, the optical constants of the surface layer being described as a mixture of 50% ZnO and 50% void (ambient) using the Bruggeman Effective Medium Approximation (BEMA). The presence of any voids or additional phases in the bulk layer, formed e.g. by addition of aluminum dopant to ZnO, is neglected, and the optical properties are considered to be homogeneous in depth. The optical constants are calculated subsequently 'point by point', that is, without assuming any dispersion model.

For thin films deposited with an aluminum precursor dosage from 0 up to 6 at.%, the photon energy at which absorption starts is increasing. The band edge E_g is associated with the energy at which $d\alpha/dE$ reaches a maximum [9]. There appears to be a linear correlation between the shift in band gap energy and aluminum doping, E_g increasing from 3.32 to 3.65 eV. This is in agreement with findings of Sernelius et al. [10], who derived a quantitative theoretical relation between band gap energy and aluminum content in the ZnO:Al films. At 7 at.% aluminum precursor dosage, a rather dramatic change in the optical film properties occurs. No further increase of the band gap is visible, and an increase of absorption in the film takes place over the entire photon range. The real and imaginary parts of the dielectric function, ε_1 and ε_2 respectively, are shown for the complete measurement range in Fig. 2. Notice that ε_1 shows an increasing negative curvature with increasing doping level at the lower end of the energy range. Through the Kramers-Kronig relationship this trend is consistent with increasing absorption in the infrared part of the spectrum.

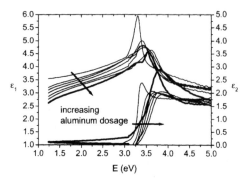

Figure 2. The dielectric function for thin ZnO:Al films with aluminum precursor dosage increasing from 0 to 7 at.% in steps of 1 at.%. The curves on the top correspond to the real part of ε (left axis), whereas the curves on the bottom correspond to the imaginary part (right axis).

From combined ellipsometric and reflection and transmission data, free charge carrier densities n_e and optical mobilities μ_{opt} are derived as parameters in the Drude oscillator model and plotted in Fig. 3. The observed trends correspond to theoretically expected trends, an increasing dopant concentration leads to an increased free electron density, but also to an increase of scattering sites, reducing the electron mobility. The effective mass of free electrons used in the calculation of these quantities is taken to be $0.28m_e$ [10]. A much higher value of $0.50m_e$ has been reported by Brehme *et al.* [11], which is explained by degeneracy combined with a possible non-parabolic behavior of the conduction band.
Notice that a higher actual effective mass results in a higher free electron density and lower electron mobility, whereas the resistivities calculated from these values remain unaffected.

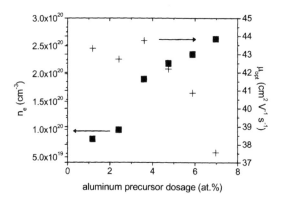

Figure 3. Electron density n_e and optical mobility μ_{opt} derived from the Drude oscillator model fit for thin ZnO:Al films with aluminum precursor dosage increasing from 1 to 7 at.%.

From these parameters obtained *thin film* resistivities are compared to the measured *thick film* resistivities in Fig. 4. The observed correspondence is explained by the difference between the DC Hall electron mobility and the optical mobility as calculated from the Drude oscillator model [12]. Electrons that are accelerated by a DC voltage may scatter either in the grain bulk (through e.g. phonon, impurity or point defect scattering) or at grain boundaries. Contrary, under the application of a rapidly oscillating electric field (such as at optical frequencies), electrons will not be displaced over a large distance but rather oscillate around an equilibrium position. Therefore, for bulk electrons, scattering will only take place by bulk mechanisms and not at grain boundaries. As for thick films the grains are relatively large, grain boundary scattering plays only a minor role compared to grain bulk scattering, and the optical and DC Hall mobilities will become close.

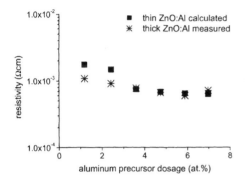

Figure 4. Measured thick film resistivities together with calculated thin film resistivities corresponding to ZnO:Al films with aluminum precursor dosage increasing from 1 to 7 at.%.

CONCLUSIONS

Textured aluminum doped ZnO films are deposited on glass substrates at a temperature of 200°C by expanding thermal plasma CVD. The addition of aluminum as a dopant results in distinct differences in film morphology, a transition from large, rounded crystallites to a more pyramid-like structure is observed. A rough surface texture, which is essential for application as front electrode in thin film solar cells, is obtained during deposition. The structure of films is hexagonal with a preferred crystal orientation in the faces (002) and (004), indicating that films are oriented with their c-axes perpendicular to the substrate plane.
Spectroscopic ellipsometry is used to evaluate both optical and electronic film properties. The presence of aluminum donors in doped films is confirmed by a shift in the ZnO band gap energy from 3.32 to 3.65 eV. In combination with reflection and transmission measurements in the visible and NIR ranges, film resistivities have been obtained from the free-carrier absorption.

These results are consistent with direct measurements. Resistivities as low as $6.0\ 10^{-4}$ Ωcm have been obtained.

ACKNOWLEDGEMENTS

The authors gratefully acknowledge Paul Sommeling (ECN) for the SEM images, Jürgen de Wolf and Miranka van den Acker (TNO TPD) for the reflection and transmission measurements, and Ries van de Sande, Jo Jansen (TU/e), Gerwin Kirchner and Leo Toonen (TNO TPD) for their skilful technical assistance. This research has been financially supported by the Netherlands Agency for Energy and Environment (NOVEM).

REFERENCES

1. R.G. Gordon, J. Proscia, F.B. Ellis, A.E. Delahoy, *Solar Energy Mat.* **18**, 263 (1989)
2. R. Groenen, J.L. Linden, H.R.M. van Lierop, D.C. Schram, A.D. Kuypers, M.C.M. van de Sanden, *Appl. Surf. Sci.* **173**, 40 (2001).
3. R. Groenen, J. Löffler, P.M. Sommeling, J.L.Linden, E.A.G. Hamers, R.E.I. Schropp, M.C.M. van de Sanden, *Thin Solid Films*, **392**, 226 (2001).
4. J. Löffler, R. Groenen, J.L. Linden, M.C.M. van de Sanden, R.E.I. Schropp, *Thin Solid Films* **392**, 315 (2001).
5. R.A. Azzam and N.M. Bashara, *Ellipsometry and Polarized Light* (Elsevier Science Publishers, New York, 1987).
6. J. Löffler, R. Groenen, E.A.G. Hamers, P.M. Sommeling, J.L. Linden, M.C.M. van de Sanden, R.E.I. Schropp, proceedings PVSEC-12, Korea (2001)
7. X.W. Sun, H.S. Kwok, *J. Appl. Phys.* **86**, 408 (1999).
8. H.G. Tompkins, W.A. McGahan, *Spectroscopic ellipsometry and reflectometry: a user's guide* (Academic Press, New York, 1999).
9. I. Hamberg, C.G. Granqvist, K.-F. Berggren, B.E. Sernelius, L. Engström, *Phys. Rev. B* **30**, 3240 (1984).
10. B.E. Sernelius, K.-F. Berggren, Z.-C. Jin, I. Hamberg, C.G. Granqvist, *Phys. Rev. B* **37**, 10244 (1988).
11. S. Brehme, F. Fenske, W. Fuhs, E. Nebauer, M. Poschenrieder, B. Selle, I. Sieber, *Thin Solid Films* **342**, 167 (1999).
12. V.I. Kayadanov, T.R. Ohno, Report NREL/SR-520-28762 (1999).

Solid Oxide Fuel Cells

Mat. Res. Soc. Symp. Proc. Vol. 730 © 2002 Materials Research Society

Micro-Fabricated Thin-Film Fuel Cells for Portable Power Requirements

Alan F. Jankowski[1], Jeffrey P. Hayes[2], R. Tim Graff[3], and Jeffrey D. Morse[3]
Lawrence Livermore National Laboratory
[1]Chemistry & Materials Science, [2]Mechanical Engineering, and [3]Electrical Engineering
Livermore, CA 94551-9900, U.S.A.

ABSTRACT

Fuel cells have gained renewed interest for applications in portable power since the energy is stored in a separate reservoir of fuel rather than as an integral part of the power source, as is the case with batteries. While miniaturized fuel cells have been demonstrated for the low power regime (1-20 Watts), numerous issues still must be resolved prior to deployment for applications as a replacement for batteries. As traditional fuel cell designs are scaled down in both power output and physical footprint, several issues impact the operation, efficiency, and overall performance of the fuel cell system. These issues include fuel storage, fuel delivery, system startup, peak power requirements, cell stacking, and thermal management. The combination of thin-film deposition and micro-machining materials offers potential advantages with respect to stack size and weight, flow field and manifold structures, fuel storage, and thermal management. The micro-fabrication technologies that enable material and fuel flexibility through a modular fuel cell platform will be described along with experimental results from both solid oxide and proton exchange membrane, thin-film fuel cells.

INTRODUCTION

The portable power source problem remains a critical issue for many consumer products as well as military objectives. While rechargeable batteries have become a hindrance in the use of consumer portable electronics, present technologies are simply inadequate for the advanced military applications of remote reconnaissance and telemetry. Therefore, new power sources are required with performance criteria specific to the direct application. A lighter-weight and longer-lasting power source provides new functionality to applications that directly benefit the end user.

Miniature fuel cells have recently received renewed interest for applications in portable power generation. Energy is stored as a fuel rather than as an integrated component of the system. Thus, fuel cells can exhibit significantly higher energy densities than batteries depending upon the type of fuel being utilized. While this has been demonstrated for power portable applications in the range of 50-500W, effective scaling of fuel cell systems has not been commercially demonstrated for power applications in the range of 1-20 W. Potential applications in this portable power range include consumer electronics as cell phones, laptop computers, video camcorders, and radios. New applications in portable power span the range of power consumption, from micro-power for long duration sensors and remote communication devices, to macro-power in the form of lighter weight sources for field use.

A miniature fuel-cell power source can be realized through an approach that combines thin film materials with micro-electro-mechanical system (MEMS) techniques.[1-5] In one idealized deployment, a micro-fabricated device (as shown in Fig. 1) consists of an integrated μ-fuel cell power source, a computation-processing unit (CPU), a telecommunication transmitter-receiver

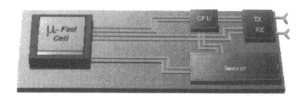

Figure 1. A micro-fabricated device is shown with an integrated μ-fuel cell power source, a computation-processing unit (CPU), telecommunication transmitter-receiver (TX-RX), and sensor.

(TX-RX), and a sensor. Based on either a solid-oxide (SOFC) or a proton-exchange membrane (PEM) electrolyte, the fuel cell device is monolithically integrated onto a manifold host structure (as shown in Fig. 2). The key structural components integral to the micro-fabricated fuel cell are the micro-machined manifold, the thermal and electrical insulating layers, the thin film electrodes, and the electrolyte layer. This architecture provides a scalable and portable fuel-cell power system for a broad range of applications that have power requirements in the range of 0.5-10 W. The MEMS approach offers a direct means to integrate the fuel cell stack with the required manifold and fuel delivery system. The MEMS integration provides a means to control the performance of the power source for specific applications.

SAMPLE PREPARATION

The use of thin film deposition with pattern and etching processes provides a direct path to the synthesis of a miniature fuel cell. The deposition sequence starts with the anode and concludes with the cathode. For a SOFC, synthesis of a 2-10 μm thin electrolyte layer is accomplished by rf-sputter deposition of an oxide target.[1] A 0.7-2.7 Pa working gas pressure of Ar is used to deposit dense and pinhole free films from a $(Y_2O_3)_6(ZrO_2)_{94}$ target. For the

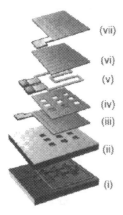

(vii)

(vi)

(v)

(iv)

(iii)

(ii)

(i)

Figure 2. A schematic of the key structural components integral to the micro-fabricated fuel cell are the (i) micro-machined base manifold, (ii) thermal and electrical insulating layer, (iii) thin film anode, (iv) electrical insulating layer, (v) heater, (vi) electrolyte layer, and (vii) thin film cathode.

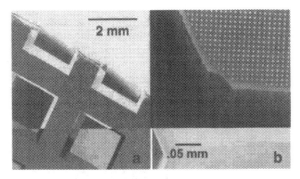

Figure 3. The use of photolithography and anisotropic etching facilitates the evolution
of decreasing dimensions in the micro-machined base manifold from (a) 2
mm x 2 mm windows, to (b) a hexagonal frame with 5 μm x 5 μm windows.

PEM fuel cell, an electrolyte membrane is formed using a spin cast process.[4] In our case, a
perflourinated sulfonic acid, i.e. Nafion™, is cast (at 5% of a 950 weight equivalent) for 30 sec
using a pipette at 500 rev-min^{-1} and then baked for 30 min at 120 °C.

Before detailing the electrode synthesis, the procedure is reviewed for preparing the micro-
machined substrate platform.[2,5] A thin layer of silicon nitride is grown using a low pressure
chemical vapor deposition process. The 0.2 μm thick film is formed at 800˚C using a 4:1 flow
ratio of dichlorosilane and NH_3 at a pressure of 33 Pa. The backside of the substrate wafer is
patterned by standard photolithographic techniques and etched to reveal individual windowed
regions of silicon nitride with areas of 4-16 mm^2 (as shown in Fig. 3a). The substrate is etched
at a typical rate of 1 μm-m^{-1} using 44% KOH at 85˚C. In a new improvement, the window size
can be reduced to improve the rigidity of the structural support for the thin film fuel cell. A two-
tiered deep anisotropic-etch procedure is developed that provides a silicon platform for the fuel
cell base electrode with channels that are less than 5 μm wide (as shown in Fig. 3b). A reactive
ion etch rate of 0.42 μm-min^{-1} is typical during subsequent silicon nitride layer removal using
500 W of power with a 2.7 Pa pressure at a 1:2 flow ratio of CHF_3 and CF_4.

The physical vapor deposition processes to produce the electrodes within the anode-
electrolyte-cathode layer stack utilize two patterning methods – photolithography and hard
masks. The sputter deposition chamber is evacuated to a base pressure less than 10^{-5} Pa. The
micro-machined silicon substrate is positioned 10 cm from an array planar magnetron sources.
A 0.6-4.0 Pa working gas pressure of Ar at a flow of 30-40 cm^3-min^{-1} is used during sputter
deposition. The deposition begins by sputtering the electrode metal target, as e.g. nickel for the
anode and silver for the cathode, at a 4-7 W-cm^{-2} forward power to yield a layer thickness of
0.5-5 μm. For photolithographic patterning, the electrode metals are deposited at room
temperature to produce a dense and fine-grained columnar structure.[3,5] After the anode
deposition, pores are formed in the layer via a photoresist mask. The pores are passages for gas
flow to the electrolyte and control the area of the catalyst-electrolyte interface. The
photolithographic pattern yields 3-5 μm diameter holes with a 3 μm intermediate spacing. After
developing the photoresist to form the template, the pattern is then etched into the Ni layer. The
etching procedure utilizes $H_2O:HNO_3$ (20:1) at 45 °C to produce an etch rate of 0.1 μm-min^{-1}.

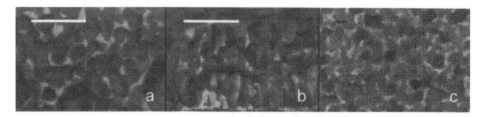

Figure 4. A porous electrode structure is imaged as sputter deposited in films of (a) nickel – in plan view, (b) nickel – in cross-section, and (c) silver – in plan view. (Bar = 1 μm.)

The thin nitride windows are then removed by reactive-ion etching to expose the anode-electrolyte interface. For the hard mask method, the electrode metal is directly deposited with an inherent continuous porosity. The hard mask is used to define the boundaries of the deposit. The as-deposited porosity is produced by sputtering the metal with a raised working gas pressure while heating the substrate to 45-55% of the melt temperature of the electrode metal. The results of this novel deposition process are shown in scanning electron micrographs for nickel and silver (in Fig. 4). The elevated temperature and gas pressure enhance both incident scattering to the substrate surface and ensuing surface mobility. This deposition condition favors recrystallization of the metal and forms bridges between the columnar structure yielding the morphology of a metallic sponge. The resultant electrode structure provides a path for electrical conductivity while the porosity enables the diffusion of reaction product species to sustain the generation of electrical current. The deposition of the electrodes sandwiches the electrolyte layer. For PEMs and SOFCs, a 5-50 nm thin catalyst layer can be deposited at the electrode-electrolyte interface.

TESTING & ANALYSIS

The PEM and SOFC are tested in similar ways. The fuel cell package to test a PEM cell (as shown in Fig. 5) consists of an integrated heater assembly, a thin-film fuel cell module, electrical pin-contacts to the electrodes, and the manifold structure for fuel supply. The major difference for the SOFC is that more refractory materials are used, i.e. stainless steel as compared to the plastic housing for the PEM cell, and that an external furnace heater can be used for high test temperatures. The fuel cell current is measured as the voltage is incrementally loaded using a semiconductor parameter analyzer. Air or an Ar-20% O_2 gas mixture is supplied to the cathode and a humidified Ar-4% H_2 gas mixture is supplied as the fuel to the anode at a 2-6 cm^3-m^{-1} flow rate. A backflow pressure up to 0.010 atm, i.e. an equivalent to 10 cm of liq. H_2O, is applied to the anode side. A thermocouple is placed in contact with the manifold structure to measure the test temperature. Results for micro-fabricated PEM and SOFC (as shown in Fig. 6) yield current- voltage-power outputs representative of the component materials as typically used in bulk designs. The PEM cell is tested at 40 °C by applying a 6 V potential to the integrated Pt heater. The open-circuit voltage (OCV) for this cell is 0.9 V. The electrodes form a 1 cm x

Figure 5. The micro-fabricated PEM and fuel cell package is shown with an integrated heater assembly, a thin-film fuel cell module, the electrical pin-contact leads, and the manifold structure.

1 cm area of contact to the polymer electrolyte. This PEM cell yields a computed peak power of 37 mW-cm^{-2} at a cell potential of 0.45 V under these test conditions. The power output remains stable after 24 hours of continuous operation. The SOFC yields an OCV of 1.0 V for the test temperatures up to 600 °C. This fuel cell is only 2.5 μm thick and is composed of a 0.5 μm thick nickel anode, a 1.2 μm thick yttria-stabilized zirconia electrolyte, and a 0.8 μm thick silver cathode. The electrodes are assumed to form a 2 mm × 2 mm contact area to the electrolyte. This SOFC yields a peak power computed to be 145 mW-cm^{-2} at a cell potential of 0.35 V under these test conditions. A summary of the computed output at intermediate temperatures is listed in Table 1. This cell was operated only once with 10 min dwell times at each test temperature.

SUMMARY

Thin film and micro-electro mechanical machining technologies are combined to micro-fabricate miniature thin-film fuel cells. A silicon modular design serves as the platform for fuel cells based on either proton-exchange or solid-oxide membranes. Packaged devices are

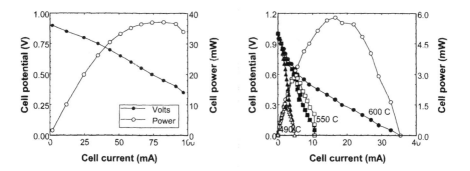

Figure 6. The current-voltage-power output of micro-fabricated PEM (left) and SOFC (right).

Table I. Peak power (mW-cm^{-2}) of a micro-fabricated SOFC at test temperature (°C)

Test Temperature (°C)	390	450	490	550	600
Cell Potential (V)	0.60	0.60	0.55	0.55	0.35
Power Ouput (mW-cm^{-2})	6	19	36	73	145

demonstrated using a hydrogen gas fuel supply with power outputs up to 0.15 W-cm^{-2}. The advent of micro-fabricated thin-film fuel cells offers a new solution for portable power requirements.

ACKNOWLEDGMENTS

This work was performed under the auspices of the U.S. Department of Energy by University of California, Lawrence Livermore National Laboratory under contract No. W-7405-Eng-48.

REFERENCES

1. A.F. Jankowski, in *Ionic and Mixed Conducting Ceramics III*, edited by T. Ramanarayanan, (Electrochem. Soc. Proc. **97-24**, Pennington, PA, 1998) pp. 106-113.

2. A.F. Jankowski and J.D. Morse, in *Materials for Electrochemical Energy Storage and Conversion II*, edited by D. Doughty, D. Ginley, B. Scrosati, T. Takamura and Z. Zhang, (Mater. Res. Soc. Symp. Proc. **496**, Pittsburgh, PA, 1998) pp. 155-158.

3. A. Jankowski, T. Graff, J. Hayes, and J. Morse, in *Solid Oxide Fuel Cells VI*, edited by S. Singhal and M. Dokiya, (Electrochem. Soc. Proc. **99-19**, Pennington, PA, 1999) pp. 932-937.

4. J. Morse, A.F. Jankowski, R.T. Graff, and J. Hayes, J. Vac. Sci. Technol. A **18**, 2003 (2000).

5. J. Morse, R. Graff, J. Hayes, and A. Jankowski, in *New Materials for Batteries and Fuel Cells*, edited by D. Doughty, H. Brack, K. Naoi, L. Nazar, (Mater. Res. Soc. Symp. Proc. **575**, Pittsburgh, PA, 2000) pp. 321-324.

Mat. Res. Soc. Symp. Proc. Vol. 730 © 2002 Materials Research Society

Grain Size and Chemical Composition Effects on the Grain Boundary Resistance of Ceria

Xiao-Dong Zhou, Harlan U. Anderson, and Wayne Huebner
Electronic Materials Applied Research Center, Department of Ceramic Engineering
University of Missouri-Rolla, Rolla, MO 65401, U. S. A.

ABSTRACT

Studies related to the effects of grain size (30nm – 5.0μm) on the electrical conductivity of undoped CeO_2 and $Ce_{0.90}Gd_{0.10}O_{1.95}$ were performed. A series of impedance spectra as a function of temperature and grain size were analyzed. It was found that the ratio of the grain boundary resistance to the total resistance became lower with decreasing grain size, increasing temperature or increasing Gd content. For the case of Gd doped CeO_2, the source of the grain boundary resistance may be due to the trapping of oxygen ions in the grain boundary area.

INTRODUCTION

Solid oxide fuel cells (SOFC) that can be operated over the intermediate temperature (IT) regime (500°C ~ 700°C) are of great importance for the commercialization of these the cells. Compared to the conventional high temperature cell operation (>800°C), the intermediate temperature operation requires an extremely strict material selection. A low resistance of the electrolyte is a key component in IT-SOFC in order to achieve a useable current density. Due to the higher oxygen ion conductivity of $Ce_{0.90}Gd_{0.10}O_{1.95}$ (0.025 $\Omega^{-1}\cdot cm^{-1}$ at 600°C) compared to zirconia-based materials (<0.005 $\Omega^{-1}\cdot cm^{-1}$), (1-3) doped CeO_2 materials are being considered as intermediate temperature (500 to 700°C) solid electrolytes for SOFC's. Moreover, operation at intermediate temperature lessons those disadvantages that CeO_2 based oxides possess at high temperature and low oxygen partial pressure, namely a significant electronic conductivity and dimensional instability due to the formation of oxygen vacancies and the associated reaction of Ce^{4+} to Ce^{3+}.

However, lower temperature operation does pose a problem due to the higher activation energy of the grain boundary resistivity, ρ_{gb}. A high ρ_{gb} can be due to many factors, including (1) amorphous or impurity phases, (2) dopant segregation, (3) an altered local defect chemistry due to space charge effects, and (4) intergranular porosity (small effect). These effects are all strongly related to grain size and the associated grain boundary area. Among these, the first factor is typically predominant, as impurities such as silicon form insulating phases that tend to wet the grains, and hence effectively block the ionic current. An inspection of the literature (4-6) concerning the dependence of the influence of oxygen activity on the electrical conductivity, suggests that the variation of impurity levels play a major role in the interpretation of the data since slopes for the log(σ) vs. log(pO_2) have been reported to vary from –1/4, -1/5 and –1/6 in the intrinsic region.

Therefore, the roles of grain size, impurities level and chemical composition have to be separated and investigated in order to understand the grain boundary resistance behavior in CeO_2. In this study, microstructural and grain boundary effects were investigated for CeO_2 and $Ce_{0.90}Gd_{0.10}O_{1.95}$. The effects of operation temperature, grain size and dopant (Gd) on the grain boundary resistance (R_{gb}) to the total resistance (R_t) ratio (R_{gb}/R_t) were measured and modeled.

EXPERIMENTAL DETAILS

Nanocrystalline CeO_2 (99.9%) based powders were synthesized using a semi-batch reactor (7), followed by heat treatment at 300°C for 1h. The powders were uniaxially pressed into pellets and bars at 150MPa, followed by isostatic pressing (310MPa) that produced a green density >50% of the theoretical value (7.1g/cm^3 for CeO_2). Various sintering temperatures were used with a heating rate of 3°C per min and a holding time of four hours at the maximum temperature. Scanning electron microscopy (SEM) was used to investigate the raw materials and sintered ceramics.

Four-point dc and two-point ac impedance spectroscopy (IS) measurements were performed using a Solartron 1260 frequency response analyzer with a 1296 interface, with an applied voltage of 0.01V over a frequency range of 1Hz to 1MHz. Thick film Ag or Pt (T > 900°C) electrodes (Electro-Science Lab, 9912-G) were used for all measurements. Data were collected for temperatures ranging between 300 and 1000°C. All electrical measurements were performed only after a stable value of the dc conductivity was achieved. Z-plot software (Scribner Associates Inc.) was used to develop appropriate equivalent circuits for modeling the impedance data.

RESULTS AND DISCUSSION

The influence of temperature and impurity

Figure 1 shows a plot of the R_{gb}/R_t ratio vs. temperature along with a plot of $\ln(\sigma T)$ vs. 10000/T analyzed from the 2-probe ac impedance spectra on the CeO_2 with an average grain size ~5μm. By assuming small polaron conduction, the plot of $\ln(\sigma T)$ vs. 1/T will give the activation energy (E_a). The calculated activation energy and the enthalpy of oxygen vacancy formation are listed in Table I. Lower activation energy of the grain ($E_{a,\,g}$ ~ 1.4eV) was observed in the measuring temperature regime, compared to the total activation energy (E_a) and the grain boundary activation energy, ($E_{a,\,gb}$) (~ 2.1eV). The difference between $E_{a,\,g}$ and $E_{a,\,gb}$ was attributed to the higher R_{gb}/R_t ratio (~ 84%) measured at 550°C than that (~ 37%) at 850°C. In this study, four-probe dc measurements, from which the effect of the electrode can be eliminated, were performed to compare the conductivity from ac impedance. The resistance achieved from dc experiment was the same as the combination of grain and grain boundary resistance from ac impedance. Therefore, the dc conductivity measurements could only reveal the total resistance in CeO_2, where the grain boundary contribution dominated at relative low temperature. Separation of the R_{gb} from total resistance is necessary in order to interpret the dependence of oxygen activity or the temperature.

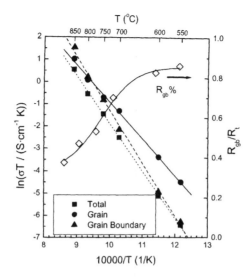

Figure 1. Plot of R_{gb}/R_t vs. temperature along with a plot of $\ln(\sigma T)$ vs. 10000/T analyzed from the 2-probe ac impedance spectra on the CeO₂ with an average grain size ~ 6.5 μm.

TABLE 1 Activation Energies for CeO₂ with various grain size measured in air

Grain Size (μm)	E_a (eV)			ΔH (eV)	
	Grain	g.b.	Total	This study	Ref.
0.4	1.7±0.1	1.6±0.1	1.7±0.1		
2.1	1.7±0.1	1.7±0.1	1.7±0.1		3.94
6.5	1.4±0.3	2.1±0.2	1.8±0.2	3.4±0.2	

The influence of grain size

Figure 2 shows the impedance spectra of CeO₂ measured in air with different grain size. Only one semicircle was observed for the finer grain size sample (< 100nm) and two successive semicircles were observed for the samples with grain size between 0.4μm and 6.5μm. Figure 3 illustrates the percentage of R_{gb} compared to the total resistance for various grain size specimens at 600°C and 800°C. The grain boundary contribution could be negligible at T>600°C when the grain size was less then 0.1μm. This grain size dependence is probably due to the presence of Si impurities located in the grain boundary area that results in the formation of blocking layers. (8)

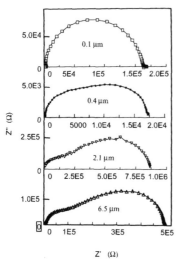

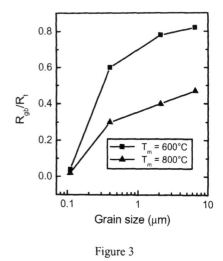

Figure 3

Figure 2

Figure 2. AC impedance spectra of low purity CeO₂ with various grain sizes, measured at 600°C in air.

Figure 3. The percentage of grain boundary over total resistance vs. grain size of CeO₂ measured at 600°C and 800°C.

The influence of dopants

As discussed previously, in addition the influence of the temperature and grain size, the presence of dopants could change the grain boundary resistance of CeO_2. Figure 4 illustrates this effect on the impedance spectra of $Ce_{0.90}Gd_{0.10}O_{1.95}$ and CeO_2 with a similar grain size (~ 0.4 μm) measured at 600°C. The contribution of the grain boundary to the total resistance was negligible for $Ce_{0.90}Gd_{0.10}O_{1.95}$ at T>500°C. Hence, the R_g determined from 2-probe ac impedance spectra was identical to obtain from the 4-probe dc measurement. On the other hand, the R_{gb} contributed around 53% of total resistance for CeO_2. The results from previous work (9) showed that the electrical conductivity related to the grains for 1.1μm $Ce_{0.90}Gd_{0.10}O_{1.95}$ was nearly identical to that for 0.15μm specimen, but that for the grain boundary is significantly higher in the micrometer grain size $Ce_{0.90}Gd_{0.10}O_{1.95}$ and decreases with decreasing grain size. These results are consistent with those found for doped ZrO_2 electrolytes (10) with a grain size ranging from 1 µm up to single crystals. In addition, in this grain size regime (from 0.15 µm to 1µm), no significant enhanced ionic conductivity of the grains was observed. Figure 5 illustrates this effect by plotting the ratio of R_{gb} to R total vs. grain sizes measured at 300, 400 and 500°C for $Ce_{0.90}Gd_{0.10}O_{1.95}$, in which higher percentage of R_{gb} was observed for finer grain size $Ce_{0.90}Gd_{0.10}O_{1.95}$, whereas in undoped CeO_2, the trend is opposite, i.e. the R_{gb}/R_t decreases with decreasing grain size. The higher R_{gb} contribution in finer grain size $Ce_{0.90}Gd_{0.10}O_{1.95}$ was very likely due to the oxygen ion trapping in the grain boundary area (9). In nanocrystalline undoped sintered CeO_2, the R_{gb} contribution is lower than that in micrometer specimen because of the presence of higher grain boundary area, a higher defect area compared to the grain, which contributes to a higher grain boundary electrical conductivity and a lower R_{gb}/R_t.

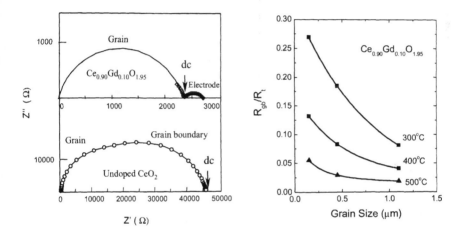

Figure 4. A comparison of impedance spectra between CeO_2 and $Ce_{0.90}Gd_{0.10}O_{1.95}$
Figure 5. Grain boundary resistance ratio of $Ce_{1.90}Gd_{0.10}O_{1.95}$ vs. grain size at the measuring temperatures of 300, 400 and 500°C.

CONCLUSIONS

Analysis of a series of impedance spectra as a function of temperature and grain size revealed that smaller grain size undoped CeO_2 (< 0.1 µm) has smaller overall grain boundary resistance than that observed in to the larger grain size (> 0.4µm). Doping with Gd increases the grain boundary conductivity and lowers the percentage of grain boundary resistance compared to the total resistance. The grain boundary activation energy ($E_{a,gb}$ ~ 2.1eV for 6.1µm CeO_2) in CeO_2 is much higher than that (1.2eV) in Gd doped CeO_2. The $E_{a, gb}$ is size dependent in CeO_2 and is independent of size for Gd-doped CeO_2, which indicates that the grain boundary resistance in CeO_2 may be due to the impurity block and in $Gd-CeO_2$ may be due to the oxygen ion trapping.

ACKNOWLEDGMENTS

This research was partially supported by the US Department of Energy under Contract DE-A26-99FT40710.

REFERENCES
1. H. Inaba, and H. Tagawa, *Solid State Ionics* **83**, 1, (1996).
2. M. Mogensen, N. M. Sammers, and G. A. Tompsett, *Solid State Ionics* **129**, 63 (2000).
3. B. C. H. Steele, *Solid State Ionics* **129**, 95 (2000).
4. E K. Chang, and R. N. Blumenthal, *J. Solid State Chem.* **72**, 330 (1988)
5. J.-H. Hwang, and T. O. Mason, *Z. Phys. Chem. (Munich)* **207**, S. 21 (1998).
6. Y.-M. Chiang, E. B. Lavik, I. Kosacki, H. L. Tuller, and J. Y. Ying, *J. of Electroceram.* **1**, 7 (1997).
7. X. D. Zhou, W. Huebner and H. U. Anderson, *Appl. Phys. Lett.* **80**, 3814 (2002)
8. C. Tian and S.-W. Chan, "Ionic Conductivities and Microstructures of CeO_2:Y_2O_3 solid electrolytes," *Mat. Res. Soc Symp. Proc.* Vol. **548**, 623; and Grain Boundary Conductivities of 0.58% Y_2O_3 Doped CeO_2 Thin Films, ibid, 629 (1999).
9. X.-D. Zhou, W. Huebner, I. Kosacki, and H. U. Anderson, *J. Am. Ceram. Soc.* **85**, 1757 (2002).
10. M. Aoki, Y.-M. Chiang, I. Kosacki, L. J. Lee, H. Tuller, and Y. Liu, *J. Am. Ceram. Soc* **79**, 1169 (1996).

Poster Session

Mat. Res. Soc. Symp. Proc. Vol. 730 © 2002 Materials Research Society   V5.3

Quantitative Structure-Activity Relationships for New Aerospace Fuels

Steven Trohalaki[1] and Ruth Pachter
Air Force Research Laboratory, Materials & Manufacturing Directorate
Wright-Patterson Air Force Base, OH 45433-7702, U.S.A.

ABSTRACT

The design of new materials can be made more efficient if toxicity screening is performed in early rather than in late stages of development, especially for volatile materials such as lubricants, fire retardants, fuels, and fuel additives. In our continuing efforts to develop methods for the prediction of the toxicological response to materials of interest to the U.S. Air Force, we have constructed Quantitative Structure-Activity Relationships (QSARs) for thirteen newly proposed propellant compounds. We employed two previously published *in vitro* toxicity endpoints in primary cultures of isolated rat hepatocytes, which measured decrease in mitochondrial function and glutathione depletion [1]. Molecular descriptors were obtained using *ab initio* molecular orbital theory. QSAR models were then derived for each endpoint. Correlation coefficients for 2- and 3-parameter QSARs exceed 0.9, possibly enabling toxicity predictions for similar compounds. Insight into the biophysical mechanism of toxic response can be gained from interpretation of the descriptors comprising the QSARs.

INTRODUCTION

Hydrazine and its methylated derivatives are powerful reducing agents with a wide range of uses, including aerospace fuels. The introduction of hydrazine, monomethyl hydrazine, and 1,1-dimethyl hydrazine as propellants grew out of a need for high-energy, noncryogenic, liquid fuels that can be used alone or mixed with other components. The toxicity of hydrazine propellants is a substantial operational concern to the U.S. Air Force as well as to the aerospace industry. The two most likely occupational exposure routes are inhalation and skin exposure [2].

Hydrazine and its derivatives enter the environment primarily from aerospace emissions and from manufacturing facilities. Toxic effects due to hydrazine exposure include liver damage, hypoglycemia, disorders of the central nervous system, interference with intermediary metabolism, induction of systemic lupus erythematosus, and carcinogenicity. Monoalkylated hydrazine derivatives are metabolically oxidized and form carbon-centered radicals [3]. Disubstituted derivatives, however, are oxidized to carbocations as well as to carbon-centered radicals [4]. Toxicity of hydrazine derivatives has been attributed to carbocations [5], carbon-centered radicals [6], and to reactive oxygen species [7].

In view of the toxicity of hydrazines, alternative propellant compositions are currently under investigation. Several molecular components were synthesized and are currently being evaluated (by the Propulsion Directorate of the Air Force Research Laboratory, Edwards AFB, CA), including those listed in Table I. These high-energy chemicals (HECs) may be categorized as hydrazine-based, amine-based, triazole-based, and a quaternary ammonium salt. A preliminary report of some of these HECs suggests that their toxicity is mediated through oxidative stress [8], a disturbance of cellular oxidation-reduction homeostasis.

In order to maintain a safe working environment, it is necessary to develop reliable, rapid and inexpensive methods for predicting health risks of newly developed chemicals. We employ

---

[1]Technical Management Concepts, Inc., P.O. Box 340345, Beavercreek, OH 45434-0345, U.S.A

here previously published *in vitro* toxicity data for these HECs to derive quantitative structure-activity relationships (QSARs) to assist in the design and optimization of chemicals for new propellant formulations.

QSARs are models that attempt to relate chemical structure to biological endpoints such as pharmacological activity or toxicity [9]. For a congeneric set of compounds assumed to act by a common mechanism, differences in chemical structure are mapped to changes in biological activity [10]. Typically, the negative log of the biological activity is expressed as a linear combination of descriptors that are calculated from chemical structure. The molecular descriptors used in toxicity prediction can be categorized as constitutional, topological, geometrical, and those derived from quantum chemical calculations, including not only molecular orbital energies but also thermodynamic and electrostatic quantities [11]. Quantum chemical descriptors are relevant to the mechanism of toxic response for simple hydrazine derivatives and are used exclusively in this work

Toxicological endpoints usually do not involve well-defined mechanisms, i.e., multiple mechanisms can lead to the same toxicity endpoint [12]. Toxicity QSARs therefore have two goals: toxicity prediction and distinguishing between possible mechanisms of toxic response, or, if mechanistic information is lacking, the formulation of the biophysical mechanism of toxic response. QSARs provide a screening capability for additional newly proposed propellants if the QSAR is valid for the new propellants in question. QSARs must be used only for interpolation and not for extrapolation, although such a distinction is not always straightforward because common mechanisms of toxic response among chemicals is not always evident.

COMPUTATIONAL METHODS

Previously [8], all HECs were modeled as neutral species (except for DMTN) employing a semi-empirical molecular orbital method, specifically, AM1 [13]. While it has been shown that somewhat better QSARs are obtained at higher levels of quantum theory [14], AM1 was a reasonable initial approach. In this study, all HECs were modeled as cations using the *ab initio* Hartree Fock method and the 6-31G** basis set (HF/6-31G**). Molecular geometries were optimized and vibrational frequencies calculated in the gas phase using Gaussian98 [15].

We employ the same data (measured in primary cultures of isolated rat hepatocytes) as used previously [1]: the effective HEC concentration resulting in a 25% reduction in mitochondrial function ($EC25_{MTT}$) and the effective concentration resulting in a 50% depletion of GSH ($EC50_{GSH}$). Only lower limits to the $EC25_{MTT}$ could be determined for four of the HECs and only lower limits to the $EC50_{GSH}$ could be ascertained for five of the HECs. The data are in Table II. Lower limit data (denoted with >) was not used to derive QSARs.

CODESSA, version 2.061 [11] was used to extract data from the Gaussian98 output files, calculate the descriptors, and perform the regression analyses required to derive the appropriate QSARs. In the so-called heuristic method, a pre-selection of descriptors occurs. Descriptors unavailable for some compounds are discarded, as are descriptors that are invariant for all compounds in the data set. Descriptors that correlate poorly are also discarded. Additional descriptors are discarded when high correlations between descriptors are found. The remaining descriptors are then ranked according to their correlation coefficients (r). Correlations of all pairs of descriptors (2-paramteter QSARs) are calculated and the best combinations are saved as working sets. The ten QSARs with the highest r and the ten with the highest F-values are then kept. Each working set is combined with the remaining descriptors to derive 3-paramteter

Table I. Proposed HECs

Chemical Name	Abbreviation	Cationic Species	Category
Hydrazinium nitrate	HZN	$[NH_2NH_3]^+$	hydrazine
2-Hydroxyethylhydrazinium nitrate	HEHN	$[NH_2NH_2CH_2CH_2OH]^+$	hydrazine
1,2-Diethylhydrazinium nitrate	DEHN	$[CH_3CH_2NHNH_2CH_2CH_3]^+$	hydrazine
Methylhydrazinium nitrate	MHN	$[CH_3NH_2NH_2]^+$	hydrazine
1,4-Dihydrazinotetrazine nitrate	DHTN		hydrazine
Diaminoguanidine nitrate	DAGN	$[H_2NC(NHNH_2)_2]^+$	hydrazine
Nitroaminoguanidine nitrate	NAGN	$[NH_2NHC(NH_2)NHNO_2]^+$	hydrazine
Ethanolamine nitrate	EAN	$[NH_3CH_2CH_2OH]^+$	amine
Histamine dinitrate	HDN		amine
Methoxylamine nitrate	MAN	$[NH_3OCH_3]^+$	amine
1,2,4-Triazole nitrate	TN		triazole
4-Amino-1,2,4-Triazole nitrate	ATN		triazole
2,2-Dimethyltriazanium nitrate	DMTN	$[(NH_2)_2N(CH_3)_2]^+$	ammonium salt

QSARs. Again, the ten QSARs with the highest r and the ten with the highest F-values are saved. This is repeated until the maximum specified order is achieved. A best multi-linear regression algorithm is also available, which obtains the best overall QSARs of a given order, i.e., the single best 2-parameter QSAR, single best 3-parameter QSAR, etc., are found.

RESULTS & DISCUSSION

Our preliminary study modeled the HECs in Table I as neutral species using a semi-empirical molecular orbital method [1]. The best 3-parameter QSARs for $EC25_{MTT}$ and $EC50_{GSH}$ that were also consistent with the lower-limit data, were [1]:

$$-\log EC25_{MTT} = 4.38 + 0.627\, E_{HOMO} - 12.4\, BO_{\sigma-\pi} - 0.390\, BO_N \qquad (1)$$

$$-\log EC50_{GSH} = -1.27 - 0.425\, \mu + 0.0162\, \alpha - 14.0\, BO_{\sigma-\pi} \qquad (2)$$

where E_{HOMO} is the energy of the highest occupied molecular orbital, $BO_{\sigma-\pi}$ is the maximum $\sigma-\pi$ bond order, BO_N is the maximum bond order of a nitrogen atom, μ is the dipole moment, and α is the polarizability (see Table III for relevant statistics). According to Koopman's theorem, the negative of the HOMO energy of a neutral species is its ionization potential (IP). This is consistent with the experimental evidence that the mechanism of toxic response, at least for hydrazine derivatives, involves oxidation. A measure of the relative ease with which an HEC is oxidized, i.e., its IP, is therefore expected to be an apt descriptor as it is in Eqn. 1. Two of the descriptors in Eqn. 2 (μ and α) are not particularly amenable to hypothesizing a biophysical mechanism, although Eqn. 2 may be useful for screening additional candidates.

The chemical species used in the toxicity studies were the nitrate salts, which, in solution, consist of the protonated form of the HEC and $[NO_3]^-$. We had argued [1] that the descriptors for the neutral and protonated forms may be similar and it is reasonable to assume that in solution the protonated and neutral forms of the HECs exist in equilibrium. If the neutral form is metabolically active and undergoes a chemical reaction, the equilibrium would be driven toward the neutral. Descriptors calculated from the neutral species are therefore appropriate. Still, it is worthwhile to see if the descriptors calculated from the cationic forms of the HECs lead to further insight into the biophysical mechanisms or to an improved predictive capability.

The majority of the best descriptors calculated using HF/6-31G** are Fukui functions, f_k^α, [16] condensed to a particular atom k [17]. Also known as local reactivity indices, f_k^α serve as a measure of the reactivity with nucleophiles ($\alpha = +$) or with electrophiles ($\alpha = -$) by considering the highest occupied and lowest unoccupied molecular orbitals (HOMO and LUMO), respectively. Both HOMO and LUMO are important for reactions with radicals ($\alpha = \cdot$).

The descriptor that best correlates with $-\log \text{EC25}_{MTT}$ is the average $f_N^\cdot$ ($f_{N,ave}^\cdot$; $r^2 = 0.661$). The average f_N^+ ($f_{N,ave}^+$) correlates second best ($r^2 = 0.497$). Another notable descriptor is the minimum valency of a nitrogen atom ($r^2 = 0.479$). The best 2- and 3-parameter QSARs for EC25_{MTT} are:

$$-\log \text{EC25}_{MTT} = -2.18 - 326\, f_{N,ave}^\cdot + 0.310\, SA_{HBA/TMSA} \qquad (3)$$

$$-\log \text{EC25}_{MTT} = -2.04 - 0.483\, f_{N,\min}^- - 292.\, f_{N,ave}^\cdot - 0.00259\, SA_{HBA} \qquad (4)$$

where $SA_{HBA/TMSA}$ is the molecular surface area that is classified as a hydrogen-bond acceptor divided by the square root of the total molecular surface area [18], and SA_{HBA} is the hydrogen-bond acceptor surface area [18]. Even though Eqn. 3 uses only two descriptors, its statistics approach those of Eqn. 1, a 3-descriptor QSAR. The statistics for Eqn. 4 are superior to Eqn. 1. The only shortcoming of Eqns. 3 and 4 is that their predictions for two of the HECs for which only lower-limit endpoints are known (DMTN and TN) are predicted to be <150 mM in contradiction with experiment (see Figure 1a).

The descriptor that best correlates with $-\log \text{EC50}_{GSH}$ is the minimum f_N^- ($r^2 = 0.802$). In addition, the maximum $f_N^\cdot$ ($f_{N,max}^\cdot$), minimum bond order for a nitrogen atom, and the LUMO energy correlate well with $-\log \text{EC50}_{GSH}$ (r^2 values range from $0.578 - 0.695$). The best 2- and 3-parameter QSARs for EC50_{GSH} are:

$$-\log \text{EC50}_{GSH} = 0.707 - 78.5\, f_{N,max}^\cdot + 3.46\, C_{min} \qquad (5)$$

$$-\log \text{EC50}_{GSH} = 2.03 - 7.22\, C_{min} - 0.108\, SA_{RPC} + 0.0188\, SA_{HBC} \qquad (6)$$

where C_{min} is the minimum net atomic charge, SA_{RPC} is the relative positive charged surface area [18] and SA_{HBC} is the hydrogen-bond charged surface area [18]. Once again, the statistics for

Table II. Toxicity Endpoints

Propellants	EC25$_{MTT}$	EC50$_{GSH}$
	(mM)	(mM)
HZN	35	20
HEHN	68	116
DEHN	50	32
MHN	18	58
DHTN	32	4
DAGN	110	145
NAGN	>150	>150
EAN	115	144
HDN	58	80
MAN	150	>150
TN	>150	>150
ATN	>150	>150
DMTN	>150	>150

Table III. Statistics for QSARs

Eqn	Tox[a]	N[b]	r^{2} [c]	F[d]	s^{2} [e]	q^{2} [f]
1	MTT	9	0.984	102.	0.0020	0.955
2	GSH	8	0.984	83.8	0.0078	0.940
3	MTT	9	0.960	72.4	0.0048	0.941
4	MTT	9	0.986	122.	0.0019	0.942
5	GSH	8	0.944	41.8	0.0226	0.879
6	GSH	8	0.996	360.	0.0018	0.976

[a] Toxic endpoint for the QSAR.

[b] N is the number of HECs used to derive the QSAR.

[c] r is the correlation coefficient.

[d] F is the Fischer criterion F value.

[e] s is the standard error.

[f] q is the cross-validated correlation coefficient (calculated using a leave-one-out strategy), which characterizes the predictive power of the QSAR.

the 2-parameter QSAR derived from cationic HECs (Eqn. 5) approach those of the 3-parameter QSAR (Eqn. 2) derived from neutral HECs (see Table III).

Eqns. 5 and 6 display the same shortcoming of Eqns. 3 and 4 – their predictions for two of the three HECs with indeterminate endpoints contradict experiment (see Figure 1b). However, Eqns. 4 and 6 are preferred over Eqns. 1 and 2 if predictions for just hydrazine-based HECs are sought because most of the indeterminate data incorrectly predicted are not hydrazines (NAGN is an exception but this may be an anomaly due to the NO_2 group). In addition, Eqn. 5 is more amenable to interpretation in terms of a biophysical mechanism of toxic response than is Eqn. 2. Moreover, the fact that the neutral-HEC QSARs (Eqns.1 and 2) are applicable to three chemical classes while the cationic-HEC QSARs (Eqns. 3-6) are only applicable to hydrazines may be evidence for different mechanisms of toxic response for these chemical classes.

Eqns. 3 and 4 are not necessarily incon
sistent with Eqn. 1. We previously hypothesized that cationic and neutral HECs exist in equilibrium [1]. When a neutral hydrazine becomes oxidized, it forms a carbon-centered free radical, which is consistent with experiment and with Eqn. 1, as discussed above. The radicals thus formed (or other radicals present in the cell) may then react with the cationic HECs, which may play a role in the HEC cation-HEC neutral equilibrium or form part of another mechanism of toxic response.

CONCLUSIONS

Improved QSARs for toxicity prediction of hydrazine-based propellants were presented. A more detailed analysis of the descriptors in Eqns. 3-6 may also lead to further insights into the biophysical mechanisms of toxic response.

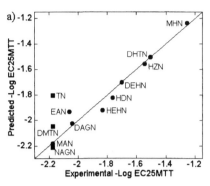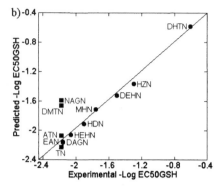

Figure 1: -Log EC25$_{\text{MTT}}$, calculated according to Eqn. 3, is plotted as a function of the experimental values in (a). Similarly, -log EC50$_{\text{GSH}}$, calculated from to Eqn. 6 is plotted in (b). The filled circles correspond to the HECs with well-defined experimental values, the squares to HECs for which only the lower-limits are known (their experimental values were taken to be the lower limits for plotting purposes). Lines drawn with a slope of unity represent a perfect correlation between theory and experiment. The predicted value for ATN was not plotted in (a) because it is off scale (-3.57). Similarly, the predicted value for MAN (-0.150) is off scale in (b).

ACKNOWLEDGMENTS

We gratefully acknowledge funding from the U.S. Air Force Office of Scientific Research.

REFERENCES

1. Trohalaki, S., Zellmer. R.J.,Pachter, R., Hussain, S.M, and Frazier, J.M. (2002) *Toxicol. Sci.*, **68**, 498
2. Keller, W.C. (1988) *Aviation, Space, and Env. Med.* **59**(11, Section 2), A100
3. Gamberini, M., Cidade, M.R., and Leite L.C.C. (1998). *Carcinogenesis*, **19**(1), 147
4. Hawks, A. and Magee, P.N. (1974) *Br. J. Cancer.* **30**, 440
5. Fiala, E.S. (1975) *Cancer*, **36**, 2407
6. Runge-Morris, M., Wu, N. and Novak, R.F. (1994) *Toxicol. Appl. Pharmacol.* **125**, 123
7. Kawanishi, S. and Yamamoto, K. (1991) *Biochemistry.* **30**, 3069
8. Hussain, S. and Frazier, J.M. (2001) *Sci. Total Environ.* **274**: 151
9. Hansch, C. and Leo, A. (1995). *Exploring QSAR: Fundamentals and Applications in Chemistry and Biology*, American Chemical Society, Washington D.C.
10. Hansch, C. and Fujita, T., (1964) *J. Am. Chem.l Soc.* **86**, 1616
11. Katritsky, A., Karelson, M., Lobanov, V.S., Dennington, R., and Keith, T., ©1994-1995
12. Benigni, R. and Richard, A.M. (1998) *A Companion to Methods in Enzymology*, **14**, 264
13. Dewar, M.J.S., Zoebisch, E.F., Stewart, J.J.P. (1985) *J. Am. Chem. Soc.* **107**, 3902
14. Trohalaki, S., Gifford, E., and Pachter, R. (2000) *Computers & Chemistry*, **24**, 421
15. Gaussian, Inc., Pittsburgh PA, 1998
16. Fukui, K. (1975) *Theory of Orientation and Stereoselection*, Springer-Verlag, Berlin
17. Contreras, R.R., Fuentealbam P., Galvan, M., Perez, P. (1999) Chem. Phys. Lett., B, 405
18. Stanton, D.T., Jurs, P.C. (1990) *Anal. Chem.* **62**, 2323

Mat. Res. Soc. Symp. Proc. Vol. 730 © 2002 Materials Research Society V5.4

Study on the doped β-FeSi$_2$ obtained by metal vapor vacuum arc ion source implantation and post-annealing

Shuangbao Wang [1]*, Hong Liang [2], Peiran Zhu [1]

[1] Institute of Physics, Chinese Academy of Sciences, Beijing 100080, PRC.
[2] Institute of Low Energy Nuclear Physics, Beijing Normal University, Beijing 100080, PRC.

ABSTRACT

β-FeSi$_2$ was firstly formed by implanting Si wafers with Fe ions at 50 kV to a dose of $5 \times 10^{17}/cm^2$ in a strong current Metal Vapor Vacuum Arc (MEVVA) implanter. Secondly, Ti implantation was performed on these Fe as-implanted samples. The Fe + Ti implanted samples were furnace annealed in vacuum at temperatures ranging from 650 to 975 °C. The XRD patterns of the annealed samples correspond to β-FeSi$_2$ structure (namely β-Fe(Ti)Si$_2$). When annealing was done above 1050 °C, the β-Fe(Ti)Si$_2$ transformed into α-Fe(Ti)Si$_2$. This implies that introducing Ti stabilizes the β-FeSi$_2$ phase. Resistance measurements were also performed.

INTRODUCTION

β-FeSi$_2$, which has an energy gap of 0.85 eV [1, 2], has been studied for it's optical and electrical properties [2, 3], as well as thermoelectric properties. Usually, Mn or Al doped β-FeSi$_2$ has p-type conduction while Co-doped FeSi$_2$ shows n-type conduction [1-3]. To improve the thermoelectric properties of β-FeSi$_2$, the doping effects of Co, Mn, Cr, V, Ti, and Cu have been studied in sintered bulk materials [16]. In fact, thin films are more suitable than sintered solids for heat-sensor applications because films are more responsive to temperature changes than bulk bodies. However, except for Mn and Co-doped films, the other additives both for p-type and for n-type β-FeSi$_2$ have not been investigated on crystalline β-FeSi$_2$ thin films [16,18].

Based on many reports [3-8], ion beam synthesis (IBS) has proven an effective method to fabricate high-quality of silicide films [3, 7, 8]. Among them, Liu et al. used an innovative MEVVA ion source implanter for which the larger current made IBS β-FeSi$_2$ films convenient [17]. Recently, we studied the MEVVA ion source IBS and found it a potential candidate to obtain high-quality β-FeSi$_2$ thin films [20].

Here, in order to know the doping influences of the implantation of additives such as Ti on β-FeSi$_2$, where its doping was known for p-type β-FeSi$_2$ [18], we did a dual implantation in the MEVVA implanter. In this experiment, we evaluated the structure and thermal stability of the doped films. One of our goals is to compare the thermal stability change of the doped β-FeSi$_2$ to the intrinsic film, which normally transforms towards α-FeSi$_2$ at about 937 °C [8,15].

EXPERIMENTAL

Flow-zone grown Si wafers (n-type <111>) were first washed by a chemical method. Fe ions were implanted at 50 kV to a fluence of $5 \times 10^{17}/cm^2$ in the MEVVA ion source implanter [9,17]. Later Ti ions were implanted at 50 kV to fluences of $1 \times 10^{16}/cm^2$ to $\sim 10 \times 10^{16}/cm^2$, respectively.

Below, we refer to the fluences by T1, T2,...,T10, where T1 is a fluence of $1\times10^{16}/cm^2$, T2 represents a fluence of $2\times10^{16}/cm^2$, and so on. The beam fluxes of Fe and Ti ions were about 60 and 42 $\mu A/cm^2$, respectively.

Furnace annealing was performed in a vacuum quartz tube oven. In the experiments, annealing was carried out over a range of 650~1100 °C, and the annealing period was 55 minutes. When annealing was finished, the samples were rapidly cooled.

X-ray diffraction (XRD) measurements were performed to identify the structure of the silicides using a D/MAX-RB diffractometer with a Cu radiation source operated at 40 kV and 80 mA. The content and profiles of metals were measured by using Rutherford backscattering spectrometry (RBS), which was performed with 2.1 MeV He$^+$ ions at a 165 degree scattering angle with an integrated current of 10 μC. Sheet resistance measurements using the four-point probe method were also performed.

RESULTS AND DISCUSSION

Fig. 1 shows XRD patterns of Fe and Fe +Ti implanted samples. In the Fe implanted samples a broad diffraction peak corresponding to the (220/202) reflection of orthorhombic β-FeSi$_2$ appeared. This peak became weaker after Ti irradiation. This implied that the as-formed β-FeSi$_2$ suffered lattice damage again by the Ti implantation. When the Fe + T10 implanted samples were annealed at 650~975 °C, except for a diffraction peak corresponding to the β-FeSi$_2$ (220/202) (see Fig. 2), there was no indication of the titanium silicide. We suggest that the as-implanted Ti atoms exist in the β-FeSi$_2$ as interstitial atoms or occupy the lattice sites of the β-FeSi$_2$. Here, the alloy is named after β-Fe(Ti)Si$_2$ because of this discrepancy. In fact, when annealing is performed at 937°C, binary β-FeSi$_2$ should transform into α-FeSi$_2$ [8,15]. However, the ternary phase appeared to be more stable in structure at even higher temperature, because the XRD pattern still showed the (220/202) reflection of the β-FeSi$_2$ at 975 °C (see Fig. 2).

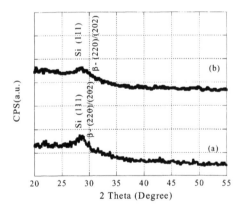
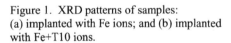
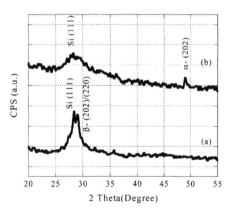

Figure 1. XRD patterns of samples:
(a) implanted with Fe ions; and (b) implanted with Fe+T10 ions.

Figure 2. XRD patterns of samples: (a) Fe+ T10 implanted and annealed at 975 °C for 55 min; (b) Fe+T10 implanted and annealed at 1050 °C for 55 min.

When the annealing temperature was increased to about 1050 °C, the diffraction line (220/202) of the β-Fe(Ti)Si$_2$ disappeared. Instead, there were the reflections of α-Fe(Ti)Si$_2$ (see Fig. 2). This showed that the alloy had changed into the tetragonal α-Fe(Ti)Si$_2$.

Fig. 3 displays the RBS spectra of the Fe implanted and the Fe + Ti implanted samples. Due to the sputtering, the Fe concentration in the Si was relatively smaller. In the Fe-implanted samples, the atom ratio of Fe to Si was 1:2.19 at its profile peak. After T10 implantation, the atom ratio decreased to 1:2.27. The iron silicide layers also became a little thinner, which implies that the dual implantation reduces the thickness of the silicide films, because the sputtering plays a larger role during the MEVVA ion source implantation.

The surface resistance of the as-implanted and the annealed samples were measured and is shown in Fig.4, where the dependences of the sheet electrical resistances on the doping fluences and annealing temperatures is indicated. At 650 °C the sheet resistance became apparently smaller (500 Ω/) comparing to the as-implanted samples (29700 Ω/). The sheet resistance continuously decreased versus temperature. When annealing was done at 950 °C, a dramatic decrement in resistance indicated the formation of the α-FeSi$_2$ [3]. However, the Fe + Ti implanted samples kept relatively larger sheet resistance even when annealed to 975 °C. The larger sheet resistance in the Fe +Ti implanted samples could be attributed to the delay in the phase transition of the β-Fe(Ti)Si$_2$ to α-phase.

Takakura et al reported [19] that β-FeSi$_2$ can be either p-type or n-type depending on Si and Fe composition ratio. A larger Si/Fe ratio corresponds to the n-type β-FeSi$_2$. In fact, our IBS β-FeSi$_2$ films were Si-rich (suggesting n-type films). Introducing Ti possibly lead to the compensation of the carriers.

Based on the experimental results above, we believe that the MEVVA ion source implantations were effective either in synthesizing or doping the β-FeSi$_2$. Due to the large beam current used, higher beam heat made the Fe atoms react to Si and nucleation occurred in the as-

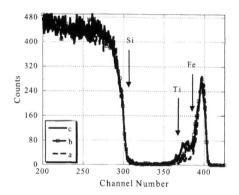

Figure 3. RBS spectra of the samples: (a) implanted with Fe + T1 ions (dash); (b) implanted with Fe + T2 ions (open squares); (c) implanted with Fe + T10 Ions (solid line).

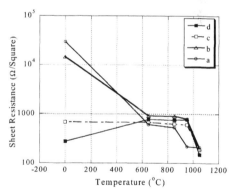

Figure 4. Change in sheet resistance of the samples versus annealing temperatures: (a) Fe implanted; (b) Fe + T1 implanted; (c) Fe + T2 implanted; (d) Fe + T10 implanted.

implanted samples [11]. The growth of the β-FeSi$_2$ was achievable through annealing. However, the doping of the β-FeSi$_2$ by the MEVVA ion source implantation was relatively rough for there were some imperfections: e.g., multi-charged ions, strong sputtering, and so on. The implantation was different from conventional implantation. However, there were also advantages, e.g., higher purity, large flux, large irradiation size, and low cost. Consequently, the MEVVA ion source IBS was more practical and more economical to synthesize and modify β-FeSi$_2$ thin films than conventional IBS. In this Fe + Ti implantation case, the IBS β-Fe(Ti)Si$_2$ was more stable in structure throughout annealing.

In general, it is understandable that the introduction of Ti enhances the thermal stability of the β-FeSi$_2$ thin films.

CONCLUSIONS

β-FeSi$_2$ films can be formed directly by MEVVA ion source implantation. Annealing is needed to get better growth. However, a disadvantage is that the Fe concentration is lower because of sputtering.

The additional Ti-implantation leads to the formation of β-Fe(Ti)Si$_2$, which is more stable at higher temperature.

ACKNOWLEDGMENT

This work is supported by the National Natural Scientific Foundation of China and Beijing City Scientific Subsidies. The authors are thankful for sample implantation at the Institute of the Low Energy Nuclear Physics of Beijing Normal University.

REFERENCES

1. D.Gerthsen, K.Radermacher, Ch. Dieker, S. Mantl, J.Appl. Phys. **71**, 3788 (1992).
2. D. Shinoda, S. Asanabe, and Y. Sasaki, J. Phys. Soc. Jpn. **19**, 269 (1964).
3. A. E. White, K. T. Short, and D. J. Eaglesham, Appl. Phys. Lett. **56**, 1260 (1990).
4. L. Ley, Y. Wang, V. Nguyen Van, S. Fisson, D. Souche, G. Vuye, and J. Rivory, Thin Solid Films **270**, 561 (1995).
5. J. P. Sullivan, R. T. Tung, and F. Schrey, J. Appl. Phys. **72**, 478 (1992).
6. D.H. Zhu, H. B. Lu, F. Pan, K. Tao, and B. Xi. Liu, J. Phys. **5**, 5505 (1993).
7. J. Desimoni, F.H. Sanchez, M.B.F. Van Raap, X. W. Lin, H. Bernas, andC. Clerc, Phys. Rev. **B54**, 12787 (1996).
8. D. J. Oostra, D. E. W. Vandenhoudt, C.W.T. Bulle-Lieuwma, E.P. Naburgh, Appl. Phys. Lett. **59**, 1737 (1991).
9. I.G. Brown, J.E. Gavin, and R.A. MacGill, Appl. Phys. Lett. **47**, 358 (1985).
10. J. P. Sullivan, R. T. Tung, and F. Schrey, J. Appl. Phys. **72**, 478 (1992).
11. Y.W. Zhang, Harry J. Whitlow, and T.H. Zhang, Opt-electronic Eng. **37/38**, 499 (1997).
12. H. Ishiwara, S. Saitoh, and K. Hikosaka, Jpn. J. Appl. Phys. **20**, 843 (1981).
13. M. C. Bost and John E. Mahan, J. Appl. Phys. **63**, 839 (1988).
14. Ingvar Engstrom and Bertil Lonnberg, J. Appl. Phys. **63**, 4476 (1988).
15. K. Radermacher, S. Mantl, Ch. Dieker, and H. Luth, Appl. Phys. Lett. **59**, 2145 (1991).
16. M. Komabayashi, K. Hijikata and Shunji Ido, Jpn. J.Appl. Phys. **30**, 331 (1991).

17. B.X.Liu, D.H.Zhu, H.B. Lu, F. Pan, and K. Tao, J.Appl. Phys. **75**, 3847-3854 (1994).
18. H.Lange, Phys. Stat. Sol. (b) **201**, 3 (1997).
19. K.I. Takakura, T. Suemasu, Y. Ikura, and F. Hasegawa, Jpn. J. Appl. Phys. **39**, L789 (2000).
20. S.B. Wang, H. Liang, Thin Solid Films (to be published).

Mat. Res. Soc. Symp. Proc. Vol. 730 © 2002 Materials Research Society

Synthesis and Thermoelectric Properties of Bi$_2$S$_3$ Nanobeads

Jiye Fang, Feng Chen, Kevin L. Stokes, Jibao He, Jinke Tang and Charles J. O'Connor
Advanced Materials Research Institute, University of New Orleans, LA 70148
E-mail: jfang1@uno.edu

ABSTRACT

Bismuth sulfide (Bi$_2$S$_3$), a direct band gap material with E$_g$ $\cong$ 1.3 eV, attracts high interest in thermoelectric investigations. In this work, nanometer-sized bismuth sulfide with unique morphology has been successfully prepared by a precipitation between bismuth 2-ethylhexanoate and thioacetamide in high-temperature organic solution with presence of proper capping/ stabilizing agents. By employing this technique, we are able to produce nanobeads of bismuth sulfide with an aspect ratio of ~ 5, typically ~10 nm wide and ~50 nm long according to the TEM observation. Characterization of XRD and TEM/HRTEM reveals that the as-prepared particles exist in single orthorhombic phase and possess high crystallinity. The composite ratio between Bi and S can be adjusted by varying the ratio between two precursors and was determined by using EDS (TEM) technique. Thermoelectric properties of these bismuth sulfide nanobeads were also investigated and will be discussed comparatively with those from commercial bulk materials.

INTRODUCTION

Bismuth sulfide (Bi$_2$S$_3$) is an important member of thermoelectric materials and is a typical semiconductor as well. In bulk form, it possesses a direct band gap of E$_g$ = 1.3 eV [1]. It is generally accepted that the band gap changes when bulk materials are transferred into nanophase due to the quantum confinement effects. It is therefore believed that the thermoelectric efficiency of nanometer-sized Bi$_2$S$_3$ may be apparently improved.

Conventionally, Bi$_2$S$_3$ nanorods can be prepared by hydrothermal method [2-4], and Bi$_2$S$_3$ nanoparticles can be obtained using microwave irradiation [5] and thermal decomposition [6,7] routes. In this work, we employed a high temperature organic solution approach and produced, for the first time, Bi$_2$S$_3$ nanobeads with controlling the aspect ratio by selection of capping agent.

EXPERIMENTAL

The chart in Figure 1 illustrates the Material synthesis. In this approach, diphenyl ether was selected as a high boiling point reaction medium. Bismuth 2-ethylhexanoate, dissolved in diphenyl ether, was used as the bismuth precursor. Thioacetamide (TAA) was used as the precursor precipitating agent. Two part systems were prepared simultaneously. For part A, TAA (99+%, 0.8744g) was added into a round bottle flask containing 1-dodecanethiol (98+%, 4 ml), di(ethylene glycol) 2-ethylhexyl ether (98%, 2 ml) and diphenyl ether (99%, 8 ml) under argon stream. The system was gradually heated to 110 °C while being stirred untill TAA was completely dissolved. For part B, oleic acid (90%, 2 ml; all above chemicals were from Aldrich) and a freshly-prepared solution of bismuth 2-ethylhexanoate (Alfa Aesar, 99.9%) in diphenyl ether (0.2M, 2 ml) were added into a three-neck flask that contains diphenyl ether (70 ml). Part A was injected into this reaction container while part B was stirred and heated to 180 °C under an Ar stream. The resulting solution was maintained at this temperature for 1 min before the heating source was removed. The mixture was then cooled down to room temperature, and Bi_2S_3 particles were precipitated by adding ethanol (60 ml) to the system and collected by centrifugation. The resulting Bi_2S_3 powder was kept in vacuum at room temperature overnight for further use. To characterize the conductivity, powder of as-prepared Bi_2S_3 was pressed into a pellet with a diameter of 9 mm and a thickness of ~1 mm. This pellet was sintered at 400 °C for 4 hours under argon atmosphere.

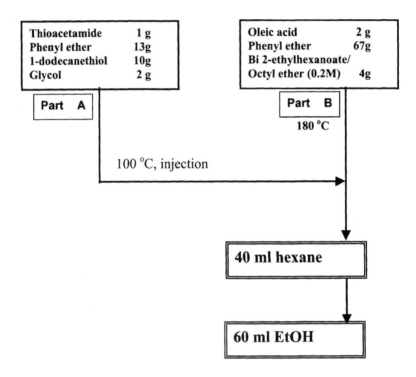

Figure 1. Flow chart for the preparation of Bi_2S_3 via high temprature solution processing.

RESULTS AND DISSCUSION

The phase identification of as-prepared Bi_2S_3 powder was performed at room temperature using a (Cu Kα radiation) X-ray diffractometer (Philips X'pert System). All the peaks in this XRD pattern, as illustrated in Figure 2, were indexed as those from orthorhombic Bi_2S_3 crystal by the comparison with the standard ICDD PDF Card of Bi_2S_3 (17-0320). Confirmed by TEM diffraction pattern (Figure 3), this XRD pattern also indicates that our product possesses high crystallinity with a single orthorhombic phase.

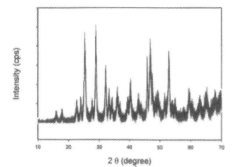

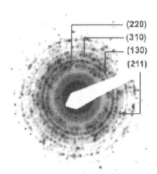

Figure 2. XRD trace of the Bi_2S_3 particles prepared at 180 °C

Figure 3. TEM image: Diffraction pattern of Bi_2S_3 particles

Figure 4 shows a light scattering result of as-prepared Bi_2S_3 particles dispersed in hexane, exhibiting the size distribution of particles. It demonstrates that the mean hydrodynamic radius of the sample is rather narrow, only ranging between 100 and 200 nm. As well known, the mean value of the size estimated using this technique is always larger than the actual one characterized by TEM due to the different instrumental nature of working function. Therefore, it is not true that both results can be always comparable. However, result from light scattering does provide the degree of particle distribution. We have also carefully checked the composition of the powder sample by using energy dispersive x-ray spectroscopy (EDS) in TEM. A typical spectrum is showed in Figure 5. The average elemental ratio between sulfur and bismuth was calculated as 58.7 to 41.3 (atomic percentage).

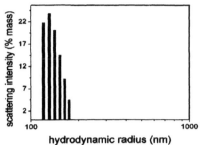

Figure 4. Bi₂S₃ particle size distribution

Figure 5 (a-c) are the transmission electron microscopy images of Bi₂S₃ particles prepared at 180 °C. Figure 5 (a) shows the morphology of the bead-shaped particles with average dimensions of ~15nm x 50nm. From our supporting experimental results, we noted that the formation of these elongated particles is most likely due to the related capping agent, i.e. 1-dodecanethiol in current investigation. Figure 5 (b) is an additional morphology image with higher magnification on selected zone, which indicates that all the elongated particles are highly crystalline. Figure 5 (c) is a high resolution TEM image showing the lattice structure on a single particle.

Figure 5. TEM images of Bi₂S₃. (a) morphology of Bi₂S₃ particles prepared at 180 °C; (b) morphology of selected Bi₂S₃ particles; and (c) HRTEM on a selected Bi₂S₃ particle

Most literature reports that Bi₂S₃ is semiconductor with a band gap of ~1.1-1.5 eV [8,9], except one research group [10], which reported metallic Bi₂S₃. We observed flat temperature dependence for resistivity (ρ) of our samples, shown as the solid curve in Figure 6. The inset of the figure was plotted in different scale to show the detailed temperature dependence of resistivity. For comparison, we also plotted the data from Ref. [10] in dashed curve. We exclude the possibility of high scattering resistance due to defects, since our data is one order of magnitude smaller than that of metallic Bi_2S_3 at room temperature [10]. It is very likely our sample is semimetal. This is supported by the fact that our Seebeck coefficient (S) is only about 1/5 of that reported by others [10]. Figure 7 shows the Seebeck coefficient plotted against temperature. It has the shape typically observed in degenerated electron gas, except the magnitude, which is a little bit smaller than most semiconductors, possibly due to the cancellation of two bands of different sign of carriers, which is the case for semimetals. Ref. [10] argued that the sulfur deficiency in Bi_2S_3 made the sample metallic. Our data showed lower resistivity than Ref. [10] and the estimated sulfur deficiency is also higher than Ref. [10]. In this case the sample should be more metallic, while we observed semimetal-like resistivity. This could be explained by either the resistivity data reflects different levels of influence of grain boundaries, or could be something special due to the nature that our raw material is nano-composite. We are currently trying different sulfur concentration and different annealing temperature to investigate this problem.

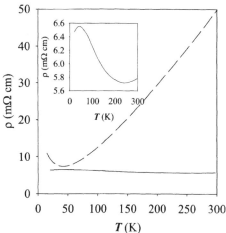

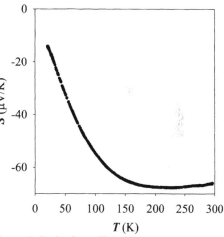

Figure 6. Comparison of resistivity of our Bi_2S_3 sample and the lowest from the literature. Solid curve is our data and the dashed curve is from Ref.[10]. Inset: same data of solid curve plotted in different scale.

Figure 7. Seebeck coefficient for our Bi_2S_3 sample.

CONCLUSIONS

Bismuth sulfide nanobeads have been successfully synthesized through a high temperature organic solution processing route. The crystallinity, particle size distribution, composite ratio and morphology of as-prepared powder were systemically analyzed using XRD, light scattering and EDS/TEM, respectively. Thermoelectric characterization indicates that our as-prepared nanostructured material is a semimetal, suggesting that sample sintered from these nanobeads of bismuth sulfide may be a promising thermoelectric material.

ACKNOWLEDGMENTS
This work was supported by DARPA through Army Research Office grant DAAD19-99-1-0001 and by LA Board of Regents, NSF/LEQSF grant (2001-04)-RII-03.

REFERENCES

[1] J. Black, E. M. Conwell, L. Seigle and C. W. Spencer, J. Phys. Chem. Solids, **2**, 240 (1957).
[2] M. W. Shao, M. S. Mo, Y. Cui, G. Chen and Y. T. Qian, J. Cryst. Growth, **233**, 799 (2001).
[3] W. Zhang, Z. Yang, X. Huang, S. Zhang, W. Yu, Y. Qian, Y. Jia, G. Zhou and L. Chen, Solid State Comm., **119**, 143 (2001).
[4] S. Yu, J. Yang, Y. Wu, Z. Han, Y. Xie and Y. Qian, Mater. Res. Bull. **33**, 1661 (1998).
[5] X. Liao, H. Wang, J. Zhu and H. Chen, Mater. Res. Bull., **36**, 2339 (2001).
[6] R. Nomura, K. Kanaya, H. Matsuda, Bull. Chem. Soc. Jpn., **62**, 939 (1989).
[7] A. Cyganski and J. Kobylecka, Thermochim. Acta, **45**, 65 (1981).
[8] R. K. Nkum, A. A. Adimado and H. Totoe, Mater. Sci. Eng. **B55** 102 (1998).
[9] Y. Ueda, A. Furuta, H. Okuda, M. Nakatake, H. Sato, H. Namatame and M. Taniguchi, J. Electron. Spectrosc. Relat. Phenom., **101-103** 677 (1999).
[10] B. Chen, C. Uher, L. Iordanidis, M. G. Kanatzidis, Chem. Mater. **9** 1655 (1997).

Mat. Res. Soc. Symp. Proc. Vol. 730 © 2002 Materials Research Society

A Grain Boundary Engineering Approach to Promote Special Boundaries in a Pb-base Alloy

D.S. Lee, H.S. Ryoo and S.K. Hwang[*]
School of Materials Science and Engineering, Inha University,
#253, Yonghyun-Dong, Nam-Gu, Incheon, 402-751, Korea
[*]Jointly appointed by the Center for Advanced Aerospace Materials,
Pohang University of Science and Technology, Pohang, Korea

ABSTRACT

A grain boundary engineering approach was employed to improve the microstructure of a commercial Pb-base alloy for better performance in automobile battery application. Through a combination of cold working, recrystallization and subsequent thermomechanical-processing, it was possible to increase the fraction of the low Σ coincidence site lattice boundaries up to 91% in addition to the substantial grain refinement. A preliminary electrochemical evaluation indicated a better corrosion resistance in the experimental material laden with the special boundaries. The high frequency of the coincidence site lattice boundaries in the specimens was interpreted in terms of the '$\Sigma3$ regeneration' model proposed in previous works.

INTRODUCTION

Pb-base alloys are extensively used for automobile batteries. Like many other secondary batteries, those made with Pb-alloy lose charge/discharge efficiency during usage, which originates from material degradation. With the recurrence of $PbO_2 \leftrightarrow PbSO_4$ reaction, a volume change occurs at the anode, which results in intergranular corrosion in the Pb-alloy [1]. The alloy design approach of adding elements such as Ca, Sn, Ag and Ba [1-4] is only partly successful in alleviating the problem and also is not readily amenable for materials recycling. Recently, a grain boundary engineering approach was proposed to improve the material properties related to the grain boundary character distribution (GBCD) [5]. Improvement of the mechanical properties as well as the electrochemical properties has been reported in Ni-base alloys [6] and Pb-base alloys [7-9] through promotion of the coincidence site lattice (CSL) boundaries. Motivated by these prior arts, the present work was designed to seek the means to maximize the frequency of the CSL boundaries in a commercial Pb-base alloy by a unique combination of the thermomechanical processing parameters.

EXPERIMENTAL PROCEDURE

A commercial Pb-0.09Ca-1.8Sn (all in wt.%) alloy for the anodic plate in automobile batteries was used as experimental material. The plate was continuously cast into a thickness of 10mm. In the initial condition, the strip cast plates were heavily cold rolled to 80% or 90% reduction and then recrystallized at 270°C/10min. Subsequent processing consisted of a light cold rolling and the identical recrystallization heat treatment. This process, termed as TMP, was repeated up to three times. The details of specimen processing schedules are shown in Table 1.

Table 1. Processing schedules for Pb-base alloy

Process	Initial condition (Strip cast+C.R.+270°C/10min.)	TMP		
		Cold rolling	Heat treatment	Cycles
80TMP3	C.R.=80%	30%		3
90R	C.R.=90%	-	270°C/10 min.	-
90TMP1	C.R.=90%	30%		1
90TMP2	C.R.=90%	30%		2

For optical microscopy, two kinds of etchants were used, ethanol : acetic acid : nitric acid = 76 : 16 : 18 in volume and acetic acid : hydro-peroxide (30%) = 75 : 25 in volume for delineation of the grain boundaries and for removing the rough surface layer of specimens, respectively. Grain boundary character distribution (GBCD) of the thermomechanically-processed specimens was analyzed with an electron back-scattered diffraction (EBSD) equipment (JEOL JSM-6300 / OPAL). To identify the CSL boundaries, the Brandon criterion [10] was adopted to determine the Σ number. Boundaries of the Σ values of 3 to 29 only were defined as the 'CSL' boundaries; those with the Σ number higher than 29 were classified as random boundaries. To evaluate the corrosion resistance, anodic polarization curves were obtained using a potentiostat/galvanostat equipment operating in the 4.8M H_2SO_4 electrolyte maintained at 25°C. The reference electrode and the counter electrode were SCE and Pt, respectively.

RESULTS

Thermomechanical processing resulted in a significant reduction of grain size. As shown in Fig. 1, the as-received specimen (as-cast into 10 mm-thick plate) showed an average grain size of about 180 μm. Heavy deformation followed by recrystallization heat treatment refined the initial microstructure so that the average grain size was reduced to the range of 25 μm to 87 μm depending on the schedule of the thermomechanical processing. The grain refinement was most apparent in specimen 90R that was heavily cold worked and recrystallized from the as-cast condition; compare Fig. 1(a) with 1(b).

During thermomechanical processing, the fraction of the special boundaries increased significantly, particularly in specimen 90TMP2, as shown in Fig. 2. The phenomenon may be discussed in two different categories. The first category is the initial condition, which should be compared with the as-received (strip cast) condition. In commercial cast Pb-alloy, the fraction of the CSL boundaries is approximately 17% [9]. In the as-cold rolled and annealed condition, 90R, this number is almost doubled. In Fig. 2(b) it is clear that most of these CSL boundaries belong primarily to the Σ3 boundaries and secondarily to the Σ9 boundaries. Distribution over other CSL boundaries is about even. The second category of CSL boundary laden microstructure is found in the TMP condition. During the thermomechanical processing, the fraction of the CSL boundaries generally increased. Furthermore, the distribution of the special boundaries showed an enhanced preferential concentration for the Σ3 and Σ9 boundaries.

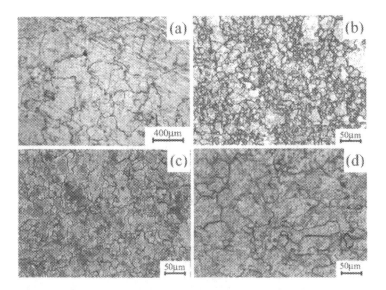

Figure 1. Optical micrographs of thermomechanically processed commercial Pb-alloy: (a) as-strip cast condition, (b) 90R, (c) 80TMP3 and (d) 90TMP2.

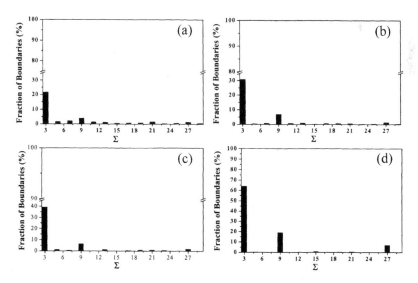

Figure 2. CSL boundary distribution of Pb-base alloy processed by various thermomechanical processing schedules (see Table 1): (a) 90R, (b) 80TMP3, (c) 90TMP1 and (d) 90TMP2.

The total fractions of the boundaries belonging to Σ3-Σ29 generated by the various thermomechanical processes are summarized in Fig. 3(a). It is clear that heavy cold working and recrystallization of the alloy in the initial condition were very effective in promoting the CSL boundaries: for example, compare the commercial material with 90R (17% vs 37%) and also compare 80TMP3 with 90TMP2 (45% vs 91%). Although there was a difference in the number of cycles in the TMP process there is a unambiguous indication that the higher the amount of the initial cold rolling, the higher is the frequency of CSL boundaries. Also note that in the present study the amount of cold rolling in the TMP process was kept low, 30%, which turned out to be crucial in multiplying the special boundaries. In summary, the present result confirms the empirical principle found by the earlier researchers [9], namely, that a heavy initial cold rolling combined with the light subsequent TMP rolling is a judicious means of promoting the CSL boundaries.

Although repeated TMP cycles continuously increased the frequency of CSL boundaries, the rate of increase was most remarkable in the first cold rolling and recrystallization; see 90R as compared to the commercial material in Fig. 3(a). This is consistent with the previous reports. For example, Kumar et al. reported that the first TMP cycle showed the highest increase of the special boundaries in OFE-Cu and the Inconel 600 alloy [11]. In the latter alloy, the fraction of the CSL boundaries increased from 30% to 70% during repeated cycles although the rate of increase diminished in the successive repetitions. About the same trend was also observed by Palumbo et al. in Pb-alloys in which the fraction was increased from 17% to 70% in repeated TMP cycles [8]. In contrast, the high fraction of the CSL boundaries obtained in the present work, 91% in specimen 90TMP2, is somewhat unusual in the sense that the rate of increment was accelerated by increasing the number of the TMP cycle from one (90TMP1) to two.

Using the OIM image of the EBSD equipment the nature of the individual special boundaries was identified, the result of which is presented in Fig. 3(b). In this specimen, 90TMP2, most of the CSL boundaries belonged to $\Sigma3^{n}$ (n=1, 2, 3). A significant portion of these boundaries, therefore, may be considered as basically the coherent twin boundaries.

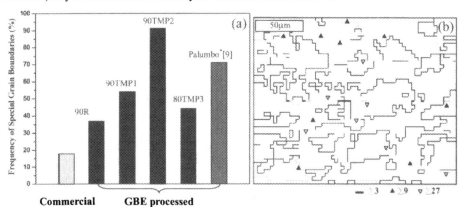

Figure 3. Analysis of the CSL boundaries in Pb alloy: (a) comparison of the total fraction of CSL boundaries (Σ3-29) generated by various combinations of cold working and subsequent TMP and (b) EBSD orientation image map of specimen 90TMP2, in which most of the special boundaries are identified as the $\Sigma3^{n}$ boundaries.

A preliminary evaluation of the corrosion resistance of the thermomechanically-processed specimens was made by the potentio-dynamic method. The anodic polarization curve shown in Fig. 4 indicates that although the corrosion potential of specimen 90TMP2 has about the same value, around –250 mV, as that of the commercial material it has less passive current density than that of the former. Thus the TMP specimen has qualitatively better corrosion resistance. This result, has been confirmed over repeated testing, is indicative of the beneficial effect of the high density of the CSL boundaries as compared to the random boundaries although further elaborate corrosion testing is required to evaluate industrial scale components.

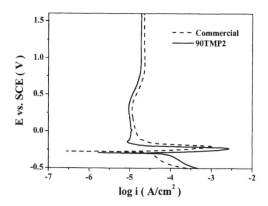

Figure 4. Anodic potentiodynamic polarization curves of Pb alloy obtained in 4.8M H_2SO_4 at 25°C at a scan rate of 1.67 mV/sec.

DISCUSSION

For metallic materials of low stacking fault energy such as Pb-base alloys there is a high possibility of increasing the special boundaries. Previous works by Palumbo et al. [8, 12-13] opened a possibility of achieving a special microstructure with abundant CSL boundaries. The present work confirms this and shows furthermore that a special microstructure with even greater population of CSL boundaries is feasible. In this approach it is recognized that a precise control of the thermomechanical processing is indispensable.

The unusually high frequency of $\Sigma3$ and $\Sigma9$ boundaries in the heavily cold worked and subsequently TMP treated specimens is considered as a result of multiplication of pre-existing $\Sigma3$ boundaries. In this respect, the '$\Sigma3$ regeneration model' proposed by Randle et al. [15-16] warrants an attention. This model is based on the hypothesis that interaction between two or more boundaries of the same character or moving random grain boundaries regenerates $\Sigma3^n$ boundaries. For example, $\Sigma9$ boundaries form at the interaction points of recrystallized grain boundaries or coherent twin boundaries ($\Sigma3_T$). Furthermore, a $\Sigma3_{GB}$ is produced by an interaction between $\Sigma9$ and $\Sigma3_T$ boundaries. Pre-existing annealing twins, therefore, play a key role in multiplying the number of $\Sigma3^n$ boundaries. Recurrent plastic work and subsequent recrystallization 'fine tunes' the boundaries within an approximate orientation range into more exact $\Sigma 3^n$ CSL orientations.

CONCLUSIONS

Based on the study of promoting special CSL boundaries in a commercial Pb-alloy for automobile battery plates, the following conclusions were drawn:

(1) A special microstructure was produced by thermomechanical processing, in which more than 91% of the entire boundaries consisted of the special CSL boundaries in addition to the concurrent substantial grain refinement, from 180 μm to about 87 μm. These special boundaries were mostly (98%) of the $\Sigma 3^n$-type boundaries.

(2) Promotion of the low Σ CSL boundaries requires an initial heavy cold working followed by the TMP treatment employing a small amount of cold rolling. The mechanism of the multiplication of the special boundaries was attributed to the 'fine tuning' of the regenerated $\Sigma 3$ boundaries as proposed in the literatures.

(3) The material processed specially to possess the high frequency of the CSL boundaries showed the electrochemical characteristic corresponding to the better corrosion resistance.

ACKNOWLEDGMENTS

This work was performed under auspices of the Korean Ministry of Science and Technology through the "2001 National Research Laboratory Program." The authors would like to thank the Global & Yuasa Battery Co., Ltd. for the supply of the experimental materials.

REFERENCES

1. A.J. Salkind, G. Mayer, David Linden, Lead-acid batteries, in: David Linden (Eds.), Handbook of Batteries and Fuel Cells, McGraw-Hill, New York, 1984.
2. E.M.L. Valeriote, J. Electrochem. Soc. 128, 1424 (1985).
3. D. Pavlov, Journal of Power Sources 48, 179 (1994).
4. L. Albert, A. Goguelin, E. Jullian, Journal of Power Sources 78, 23 (1999).
5. T. Watanabe, Res. Mechanica 11, 47 (1984).
6. P. Lin, G. Palumbo, U. Erb, K.T. Aust, Scripta Mater. 33, 1387 (1995).
7. E.M. Lehockey, G. Palumbo, P. Lin, A.M. Brennenstuhl, Scripta Mater. 36, 1211 (1997).
8. E.M. Lehockey, D. Limoges, G. Palumbo, J. Sklarchuk, K. Tomantschger, Journal of Power Sources 78, 79 (1999).
9. G. Palumbo, WO Patent No. 01/26171 A1 (12 April 2001).
10. D.G. Brandon, Acta Metall. 14, 1479 (1964).
11. Mukul Kumar, Wayne E. King and A.J. Schwarts, Acta Mater. 48, 2081 (2000).
12. M.L. Kronberg, F.H. Wilson, Trans. AIME 185, 501 (1949).
13. K.T. Aust, J.W. Rutter, Trans. AIME 215, 119 (1959).
14. H. Kokawa, T. Watanabe, S. Karashima, J. Mater. Sci. 18, 1183 (1983).
15. C.B. Thomson, V. Randle, Acta Mater. 45, 4909 (1997).
16. V. Randle, Acta Mater. 47, 4187 (1999).

Mat. Res. Soc. Symp. Proc. Vol. 730 © 2002 Materials Research Society   V5.7

Thermoelectric Properties of the Semiconducting Antimonide-Telluride $Mo_3Sb_{5-x}Te_{2+x}$

Enkhtsetseg Dashjav and Holger Kleinke*
Department of Chemistry, University of Waterloo,
Waterloo, Ontario, Canada N2L 3G1
E-mail: kleinke@uwaterloo.ca

ABSTRACT

Typically, useful thermoelectrics are small-gap semiconductors. Mo_3Sb_7 would be an interesting candidate, if it were not metallic. Electronic structure calculations reveal that metallic Mo_3Sb_7 can be made semiconducting by heavy doping, e.g., by replacing Sb in part with Te. We succeeded in the preparation of semiconducting $Mo_3Sb_{5-x}Te_{2+x}$ with enhanced thermoelectric properties. Furthermore, we incorporated small M atoms into the cubic Sb/Te cage in an attempt to create the rattling effect as found in the filled skutterudites that have attracted wide interest for their outstanding thermoelectric properties.

INTRODUCTION

Thermoelectric materials can convert heat into electricity and vice versa. This fascinating energy conversion is commercially in use, but due to its low efficiency restricted to niche technologies, such as small-scale refrigeration or power generations in remote locations (e.g. in spacecrafts, subsea, the Rockies, …). Thermoelectric devices are usually comprised of semiconductors [1]. Their performance is indicated by the *figure-of-merit* ZT, which is defined as $ZT = TS^2\sigma/\kappa$. Here, T is the temperature, S the Seebeck coefficient (thermopower), and σ and κ are the electrical and the thermal conductivities, respectively. The commercially used materials such as Bi_2Te_3 may exhibit ZT values around 1 at the ideal operating temperature; the higher ZT, the better the thermoelectric performance [2].

In the past years, the filled skutterudites [3] - among other materials such as β-Zn_4Sb_3 [4] and $CsBi_4Te_6$ [5] - have attracted wide interest because of their outstanding thermoelectric properties, which were described in 1996 [6]. Many investigations into this structure family followed subsequently [7-9]. The general formula is $Ln_xM_4Sb_{12}$ with $x \le 1$, where Ln is a lanthanoid and M a valence-electron rich transition element such as Fe, Co, Ni, … While the parent compound, $LaFe_4Sb_{12}$, is metallic, $LaFe_3CoSb_{12}$ exhibits excellent thermoelectric properties based on its experimentally determined figure-of-merit ZT, which may become as high as 1.4 at 730 °C, for its good thermopower and electrical conductivity are combined with an extraordinarily low (thus perfect) thermal conductivity. The latter stems from the high vibrations of the La atom situated in a large "cage" of Sb atoms, a phenomenon usually referred to as *rattling*.

In the past, we investigated valence-electron poor transition metal antimonides [10,11], one direction being the thermoelectric energy conversion [12,13]. In some sense, the antimonide Mo_3Sb_7 [14] is a material similar to the skutterudites: its structure contains voids that might be filled with (small) cations, and its band structure comprises a band gap of the right magnitude. Its cubic symmetry is also an advantage, for the thermopower depends on the effective band masses as well as the degeneracy of the band extrema around the Fermi level [15,16]. This was our motivation to start optimizing the properties of Mo_3Sb_7, as discussed in this contribution.

EXPERIMENTS

Syntheses. Mo_3Sb_7 can be prepared by reacting the elements in sealed silica tubes at 750 °C, which is above the melting point of antimony (630 °C). We used the same approach to prepare the mixed antimonide-tellurides $Mo_3Sb_{5-x}Te_{2+x}$ [17]. This resulted in single-phase samples for $0 \leq x \leq 2.3$, if the annealing time exceeded ten days.

For the (formal) intercalation of small M atoms (like Cu and Mg), we added these in their elemental form as well. We chose the ratio of 1 M : 6 Mo : 14 Sb, since 2 cubic voids, 12 Mo atoms, and 12 Sb(1) and 16 Sb(2) atoms occur per unit cell. To determine the actual extend to which the voids may become filled, we performed an X-ray single crystal study on $Cu_xMo_3Sb_7$ using a Smart Apex diffractometer (BRUKER, Madison, WI). The cubic void exhibited an electron density of 10.5 electrons/$Å^3$. Refining this as a Cu site resulted in a small occupancy of 8.2(1.4) % Cu, yielding $Cu_{0.041(7)}Mo_3Sb_7$. This occupancy is definitely significant, since the structure refinement of the binary Mo_3Sb_7 resulted in a featureless difference Fourier map, i.e. no noteworthy electron density existed at that position. Similarly, we found an intercalation of Mg atoms corresponding to the refined formula of $Mg_{0.07(1)}Mo_3Sb_7$.

Calculations. Self-consistent linear muffin tin orbitals (LMTO) calculations [18-20] were performed on Mo_3Sb_7, $Mo_3Sb_5Te_2$, and $Mg_xMo_3Sb_7$. Therein, the density functional theory is utilized with the local density approximation (LDA).

Physical property measurements. The Seebeck coefficients and electrical conductivities were determined on cold-pressed bars of the dimensions $5 \times 1 \times 1$ mm (using a force of 10 kN) under dynamic vacuum of 10^{-3} mbar, using the commercial device from MMR Technologies for the Seebeck measurements and a self-built device for the conductivity measurements.

RESULTS AND DISCUSSION

Crystal structure. Mo_3Sb_7 forms the Ir_3Ge_7 type comprising a cubic body-centered unit cell of space group $Im\bar{3}m$ (left part of Figure 1). The structure contains a three-dimensional Sb atom network, which includes the Mo atoms in square antiprismatic voids.

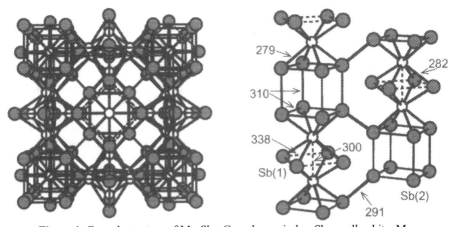

Figure 1. Crystal structure of Mo_3Sb_7. Gray, large circles: Sb, small, white: Mo.

The antiprisms build chains running parallel to each of the crystallographic axes (which are equivalent in the cubic crystal system) by forming intermediate Sb–Sb bonds between each pair of MoSb$_8$ antiprisms. This results in the formation of empty Sb(2)$_8$ cubes alternating with pairs of MoSb$_8$ antiprisms along each axis. Parallel running chains are interconnected via the shortest Sb–Sb bond of this structure between the Sb(2)$_8$ cubes (right part of Figure 1).

Intercalating a cation M into the empty cubes depicted on the right would lead to M–Sb(2) distances of 269 pm and M–Mo distances of 328 pm - reasonable distances for smaller cations such as Cu$^+$ or Mg^{2+}. It is thus not surprising that we succeeded in synthesizing Cu$_x$Mo$_3$Sb$_7$ and Mg$_x$Mo$_3$Sb$_7$, with the additional atoms being situated in the centers of these cubes.

Electronic structure. The band structure of Mo$_3$Sb$_7$, shown in Figure 2, contains a band gap above the Fermi level (fixed at 0 eV), which may be reached by adding two valence electrons. Since several bands cross the Fermi level, E$_F$, along all symmetry lines selected (see the Brillouin zone on the right [21]), Mo$_3$Sb$_7$ will exhibit three-dimensional metallic properties. The bands with mostly Sb p orbital character are emphasized via the fat band representation [22]. With respect to thermoelectric properties, it is important to note the band gap that is located two valence-electrons per formula unit above E$_F$, as well as the presence of numerous flat bands in the area around E$_F$, which indicate high effective band masses that favor high thermopower.

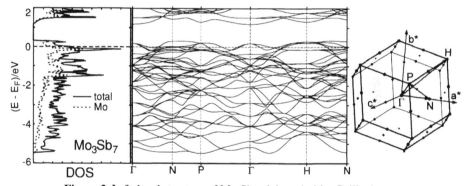

Figure 2. Left: band structure of Mo$_3$Sb$_7$; right: primitive Brillouin zone.

A band gap of ca. 0.9 eV occurs above the Fermi level. A smaller gap, namely of the size of a few tenths of an eV (i.e., 6 - 10 k$_B$T with k$_B$ = Boltzman constant, T = temperature) is recommended for enhanced thermoelectric energy conversion [23]. The region around E$_F$ is dominated by Mo d states, while most of Sb contributions are located well below and partly above E$_F$.

Two major differences between the Mo$_3$Sb$_7$ band structure and the band structure of our model for Mo$_3$Sb$_5$Te$_2$ (Figure 3) are that the Fermi level falls right into the band gap, as anticipated, and that the size of the band gap decreased from 0.9 eV to 0.5 eV. Similarly, adding Mg atoms into the cubic void leads to a decrease in the band gap as well, as illustrated in Figure

3. In that case, the Sb : Te ratio will have to be adjusted accordingly to raise the Fermi level into the band gap.

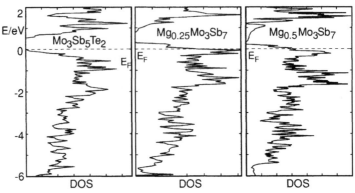

Figure 3. Densities of states for $Mo_3Sb_5Te_2$ (left), $Mg_{0.25}Mo_3Sb_7$ (middle), and $Mg_{0.5}Mo_3Sb_7$ (right).

Physical properties. The experimental proof that replacing 2 of the 7 Sb atoms per formula unit occurs with a change from metallic to semiconducting properties is shown in Figure 4. The resistivity of $Mo_3Sb_5Te_2$ is most significantly higher than that of Mo_3Sb_7, and it has the negative temperature dependence typical for semiconductors. Here it should be noted that the absolute values are severely influenced by the grain boundary effect, since cold-pressed pellets were used. It is assumed that the materials' actual resistivities are significantly lower than indicated by these measurements.

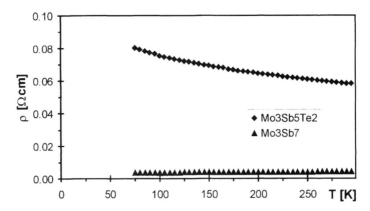

Figure 4. Electrical resistivities of Mo_3Sb_7 and $Mo_3Sb_5Te_2$.

According to our experiments, up to 2.3 Sb atoms of Mo_3Sb_7 may be replaced by Te atoms. The thermopower (Seebeck coefficient) S of the series $Mo_3Sb_{7-x}Te_x$ is drawn versus x in Figure 5. In each case, S increases with increasing temperature. Also, the thermopower S increases in the series $Mo_3Sb_{7-x}Te_x$ with increasing Te content (thus increasing valence-electron concentration), until the maximum is reached at x = 2.2 with, e.g., S = +93 at 300 K and +190 μV/K at 600 K. These values are comparable to the thermopower of $LaFe_3CoSb_{12}$, which is about +100 at 300 K and below +200 μV/K at 600 K.

Theoretically, even higher Te contents should eventually lead to negative thermopower, since at some point the electrons should dominate as the charge carriers. This point, however, could not be reached experimentally, for increasing the Te content led to the formation of $MoTe_2$ in addition to $Mo_3Sb_{7-x}Te_x$. Hypothetically, the crossover between positive and negative thermopower would occur at x = 2. Since this is not observed below x = 2.3, it is concluded that $Mo_3Sb_5Te_2$ is a semiconductor with intrinsic p type character. The same is true for $LaFe_{4-x}Co_xSb_{12}$ (p type at x = 1) and $Bi_{2+x}Te_{3-x}$ (n type at x = 0). In the latter case, the transition between p and n type takes place at 63 at.-% Te (hypothetical value: 66.7 % Te) [24].

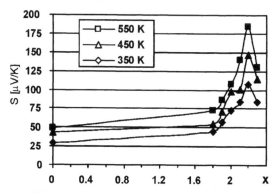

Figure 5. Seebeck coefficients of $Mo_3Sb_{7-x}Te_x$.

CONCLUSIONS

We made Mo_3Sb_7 semiconducting by a partial Sb/Te replacement. $Mo_3Sb_{7-x}Te_x$ exists with $0 \leq x \leq 2.3$. The thermopower reaches a maximum at x = 2.2, which is comparable to the values found in the ideal filled skutterudites (e.g., ca. 100 μV/K at room temperature). Since increasing temperature occurs with increasing thermopower, the material might be more interesting for high temperature applications. The two other key properties of the thermoelectric energy conversion, however, remain to be determined in detail for more accurate results.

It is possible to intercalate small cations into the cubic void of Mo_3Sb_7. This may help in lowering the phonon contribution to the thermal conductivity, thus increasing ZT and thereby further enhancing the thermoelectric properties. Incorporating Mg atoms also leads to a desired decrease of the band gap.

ACKNOWLEDGMENTS

Financial support from the *Canadian Foundation for Innovation*, the *Canada Research Chair Secretariat*, the *Ontario Innovation Trust*, the *Materials Manufacturing Ontario*, the *Province of Ontario*, and the *National Sciences and Engineering Research Council of Canada* is appreciated.

REFERENCES

1. D. M. Rowe, *CRC Handbook of THERMOELECTRICS*, CRC Press, Inc., Boca Raton, 1995.
2. D. M. Rowe and C. M. Bahandri, *Modern Thermoelectrics*, Holt Saunders, London, 1983.
3. W. Jeitschko and D. Braun, *Acta Crystallogr.* **B33**, 7 (1977).
4. T. Caillat, J.-P. Fleurial, and A. Borshchevsky, *J. Phys. Chem. Solids* **58**, 1119 (1997).
5. D.Y. Chung, T. Hogan, P. Brazis, M. Rocci-Lane, C. Kannewurf, M. Bastea, C. Uher, and M. G. Kanatzidis, *Science* **287**, 1024 (2000).
6. B. C. Sales, D. Mandrus, and R. K. Williams, *Science* **272**, 1325 (1996).
7. G. S. Nolas, D. T. Morelli, and T. M. Tritt, *Annu. Rev. Mater. Sci.* **29**, 89 (1999).
8. G. S. Nolas, M. Kaeser, R. T. Littleton IV, and T. M. Tritt, *Appl. Phys. Lett.* **77**, 1855 (2000).
9. N. R. Dilley, E. D. Bauer, M. B. Maple, S. Dordevic, D. N. Basov, F. Freibert, T. W. Darling, A. Migliori, B. C. Chakoumakos, and B. C. Sales, *Phys. Rev.* **B61**, 4608 (2000).
10. H. Kleinke, *Chem. Soc. Rev.* **29**, 411 (2000).
11. H. Kleinke, *J. Am. Chem. Soc.* **122**, 853 (2000).
12. H. Kleinke, *Inorg. Chem.* **40**, 95 (2001).
13. I. Elder, C.-S. Lee, and H. Kleinke, *Inorg. Chem.* **41**, 538 (2002).
14. A. Brown, Nature, 1965, 206, 502.
15. R. P. Chasmar and R. Stratton, *J. Electron Control* **7**, 52 (1959).
16. F. J. DiSalvo, *Science* **285**, 703 (1999).
17. E. Dashjav, A. Szczepenowska, and H. Kleinke, *J. Mater. Chem.* **12**, 345 (2002).
18. L. Hedin and B. I. Lundqvist, *J. Phys.* **C4**, 2064 (1971).
19. O. K. Andersen, *Phys. Rev.* **B12**, 3060 (1975).
20. H. L. Skriver, *The LMTO Method*, Springer, Berlin, 1984.
21. A. Kokalj, *J. Mol. Graph. Model.* **17**, 176 (1999).
22. O. Jepsen and O. K. Andersen, *Z. Phys.* **97**, 25 (1995).
23. J. O. Sofo and G. D. Mahan, *Phys. Rev.* **B49**, 4565 (1994).
24. J. P. Fleurial, J. Gailliard, R. Triboulet, H. Scherrer, and S. Scherrer, *J. Phys. Chem. Solids* **49**, 1237 (1988).

Mat. Res. Soc. Symp. Proc. Vol. 730 © 2002 Materials Research Society

A promising new photo-catalyst InVO$_4$ for water molecule decomposition in the visible wavelength region

M. Oshikiri[1,2], A. P. Seitsonen[2], M. Parrinello[2], M. Boero[3], J. Ye[4] and Z. Zou[5]
[1]Nanomaterials Laboratory, National Institute for Materials Science,
3-13 Sakura, Tsukuba, Ibaraki 305-0003, Japan
[2]Swiss Center for Scientific Computing,
Galleria 2, Via Cantonale, CH-6928 Manno, Switzerland
[3]Institute of Physics, University of Tsukuba,
1-1-1 Tennodai, Tsukuba, Ibaraki 305-8571, Japan
[4]Materials Engineering Laboratory, National Institute for Materials Science,
1-2-1 Sengen, Tsukuba, Ibaraki 305-0047, Japan
[5]Photoreaction Control Research Center,
National Institute of Advanced Industrial Science and Technology,
1-1-1 Higashi, Tsukuba, Ibaraki 305-8565, Japan

ABSTRACT

A promising vanadium-based photo-catalyst (InVO$_4$) has been proposed recently by our group. This catalyst has been shown to be active in the visible light range, up to a wavelength of about 600 nm, and is able to promote the decomposition of water. In this paper, we focus on its electronic structure, computed via first principles calculations, in order to figure out analogies and differences with similar systems (InNbO$_4$, InTaO$_4$ and BiVO$_4$) that have already been reported to act as photo-catalysts. An attempt is made to address the problem of the wavelength dependency of the photo catalytic activity of the InMO$_4$ (M = V, Nb, Ta) family and how this is related to the crystal structure. Finally, by using an *ab initio* molecular dynamics approach, we inspect the relaxation/reconstruction phenomena occurring at the exposed surface of the InVO$_4$ catalyst induced by the absorption of a water molecule, that represents a crucial step in the catalysis reaction.

INTRODUCTION

It is well known that a photo-catalyst able to promote the hydrolysis of water molecules and operating in the visible and infra-red region is highly desired for solar energy storing. If we consider the redox level difference of H$^+$/H$_2$ and O$_2$/H$_2$O, which is about 1.2 eV (1.0 μm), the problem of finding a material with such properties does not seem a tough one: even an infra-red-active semiconductor catalyst should exist *a priori*. However, this naïve picture is in contrast with the experimental evidence: to date, most of the photo catalyst can work only in the UV region (< 420 nm). In an attempt to shed some light, we inspected the electronic structure of the most common photo-catalysts. As discussed later, the conduction band bottom (CBB) of the majority of the photo-catalysts consists of d orbitals of the transition metals (e.g. Zr, Ta, Nb, Ti) present in the structure, while the valence band top (VBT) is spanned by the 2p orbitals of the oxygen atoms. Given this scenario, an intuitive way to activate the catalyst in a longer wavelength range would be to substitute vanadium to either Zr, Ta, Nb or Ti. This idea is suggested by the lower position, on the energy axis, of the atomic 3d energy levels of V with respect to the other transition metals. Provided that this downward shift holds also in the crystal, as expected, its effect would result in a lowering of the bottom of the d band, hence making it closer to the redox level of H$^+$/H$_2$. A previous work from our group has evidenced that InTaO$_4$ and InNbO$_4$ display a catalytic activity in the visible region up to a wavelength of 500 nm [1]. This has encouraged us to inspect V-based systems with some encouraging results [2]. It is worthy of note the fact that, in our experiments on hydrogen evolution, InVO$_4$ has been observed to be photo-catalytically active in a wavelength range from the UV region up to 600 nm.

In this paper, we discuss the main features of the electronic structure relevant to InVO$_4$, in comparison with those of InNbO$_4$ and InTaO$_4$, which have already been reported to be photo-catalysts. The electronic structure of BiVO$_4$ is also included in an attempt to achieve a more systematic understanding of the whole

137

photo-catalysts family reaction, since $BiVO_4$ is known to be active for oxygen evolution in the visible range [3]. The wavelength dependency of the photo-catalytic activity shown by $InMO_4$ compounds (M = V, Nb, Ta) is discussed and related to their crystal structure. Finally, some preliminary results, obtained from first principles molecular dynamics (MD) simulations, are reported; these show how relaxation and/or reconstruction processes occur at the exposed surface of the $InVO_4$ system when a water molecule is absorbed. This is an issue of a certain relevance, since it represents a crucial step in the catalytic reaction.

CRYSTAL STRUCTURES

The details concerning the synthesis of $InVO_4$, $InNbO_4$ and $InTaO_4$ have already been reported in references [2], [3] and [4] respectively. The space group and the crystal structure parameters of $InVO_4$ [5], $InNbO_4$ [4], $InTaO_4$ [4] and $BiVO_4$ [6] are summarized in Table I, while the atomic crystallographic positions for $InVO_4$, $InNbO_4$ and $InTaO_4$ can be found in references [4] and [5]. Unfortunately, the atomic positions for $BiVO_4$ have never been reported in the literature to the best of our knowledge. In an attempt to overcome this lack of data, we determined theoretically these atomic positions via density functional theory (DFT) calculations within the local density approximation (LDA). To this aim, we first fixed the (known) experimental lattice constants [6] (see Table I) and then minimized the total energy of the system. For simplification, we approximated the angle β to 90° in the present work. The computed atomic positions will be reported in a forthcoming publication [7]. For the ongoing discussion, we can anticipate that the $BiVO_4$ crystal structure can be regarded as a sort of deformed scheelite structure [8]. Indeed, the total energy functional was minimized by varying the x, y, z parameters of reference 8.

The crystal structure of $InVO_4$ is composed of two kinds of polyhedra: one is the InO_6 octahedron and the other one is the VO_4 tetrahedron. The InO_6 octahedron is connected to the next neighboring one by a shared edge. The octahedral subunits form chains along the [001] direction and the chains are linked to each other by the VO_4 tetrahedra. The six O atoms surrounding the In site are located at a distance of 2.16 Å; however, due to angular distortions, InO_6 is not a regular octahedron. In the VO_4 tetrahedron, the four O atoms are located at different distances; namely, two of them are at 1.66 Å from the V site, while the other two are at 1.79 Å. The remarkable feature of the $InVO_4$ structure is that the V-V distance is as large as 4.05 Å (while, the In-In distance is ~3.3 Å). On the other hand, only octahedral structures are present in $InNbO_4$ and $InTaO_4$: the InO_6 and NbO_6 in the former case and InO_6 and TaO_6 in the latter. Each InO_6 octahedron is connected to the next one by shared edges and the result is a zigzag chain. The chain-to-chain connection is provided by the NbO_6 (or TaO_6) octahedra to build up the three dimensional network. The distances of Ta-Ta, Nb-Nb and In-In in the cases of $InNbO_4$ and $InTaO_4$ are ~3.3 Å. In the system of $BiVO_4$, the distances of Bi-Bi and V-V are ~3.9 Å. The V site in the $BiVO_4$ system is surrounded by four O atoms as well, forming a VO_4 tetrahedron with V-O distances of 1.86-1.87 Å.

ELECTRONIC STRUCTURE

The electronic structure of the bulk $InVO_4$ system has been investigated via DFT based first principles calculations within LDA and compared with those of $InNbO_4$, $InTaO_4$ and $BiVO_4$ in order to achieve a comprehensive picture. In all the calculations presented here, we adopted the refined experimental lattice constants as determined by our experiments. These are very close to the data reported in Table I. As far as the

Table I. Experimental lattice constants (in Å) of $InVO_4$, $InNbO_4$, $InTaO_4$ and $BiVO_4$.

Catalyst	Space group	a	b	C	β
$InVO_4$	Cmcm	5.7542	8.5229	6.5797	
$InNbO_4$	P2/a	5.142	5.747	4.818	91.13°
$InTaO_4$	P2/a	5.1508	5.770	4.821	91.35°
$BiVO_4$	I2/a	5.186	11.708	5.100	90.43°

Table II. Projected weights of wavefunctions at the conduction band bottom and the valence

Catalyst	Conduction band bottom		Valence band top	
InVO₄	V_3d 59 %	In_5s 19 %	O_2p 81 %	In_5p 6 %
InNbO₄	Nb_4d 51 %	In_5s 19 %	O_2p 83 %	In_4d 5 %
InTaO₄	Ta_5d 43 %	In_5s 22 %	O_2p 81 %	In_4d 6 %
BiVO₄	V_3d 79 %	O_2p 9 %	O_2p 64 %	Bi_6s 18 %

atomic positions are concerned, also in this case we used the data provided by our own experiments. As mentioned above, this is not true for the case of BiVO₄, where we had to rely on the theoretical guess. The calculations were performed by DFT-LDA using the linear muffin tin orbital (LMTO) basis set, within the standard atomic sphere approximation (ASA) [9]. The (valence) electronic configurations used for the various atoms were: 4d, 5s, 5p for In, 3d, 4s for V, 4d, 5s for Nb, 4f, 5d, 6s for Ta, 6p, 6s for Bi and 2s, 2p for O. A spin-unpolarized scheme was adopted and the number of empty states included was chosen large enough to avoid any spurious effect due to the numerical truncation.

The projected density of state (DOS) of InVO₄, InNbO₄, InTaO₄ and BiVO₄, as obtained by the present approach, is reported in Figure 1. It is clear that the CBBs of these systems are constituted mainly by the *d* orbitals of the transition metals and the VBTs are spanned mostly by the 2p orbitals of the oxygen atoms. It can be noticed that the level of the CBBs shifts towards lower values, along the energy axis, for the three systems InTaO₄, InNbO₄ and InVO₄. The latter case corresponds to the larger downward shift and turns out to be rather similar to the case of BiVO₄. The band gaps are 3.7 eV, 3.4 eV, 3.1 eV and 1.3 eV, respectively. The measured band gaps, obtained from experiments of spectral transmittance of InTaO₄, InNbO₄ and InVO₄, read 2.6 eV, 2.5 eV and 2.0 eV, respectively. Since the samples are not single crystals, the experimental transmissivity would include some effects due to defects or impurities that might affect the energy gap. On the theoretical front, it is well known that the LDA approximation generally does not give accurate band gap estimations [10]. However, the experimental trend matches remarkably well to the theoretical result, thus providing a solid ground for the present analysis.

We projected the computed wavefunctions on the single (atomic) angular momentum components, in order to make the statement about the nature of the orbitals more quantitative. The results are shown in Table II. It can be inferred that, indeed the *d* orbitals of the transition metals are responsible for the major

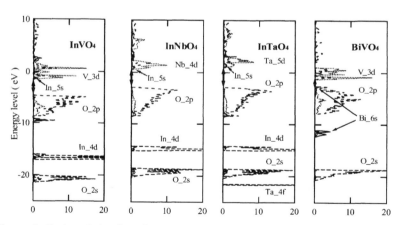

Figure 1. Projected density of state spectra of InVO₄, InNbO₄, InTaO₄ and BiVO₄ by DFT-LDA with LMTO-ASA. The band gap is indicated by arrows in the energy axis.

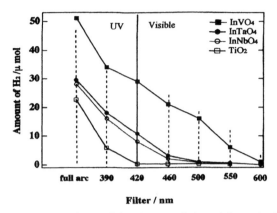

Figure 2. Wavelength dependence of the photo catalytic activity under light irradiation from full arc up to 600 nm.

contribution to the band, but also the $5s$ state of In has a large projection and hence a non-negligible weight in the CBBs of $InMO_4$ (M = V, Nb, Ta) systems. Analogously, while the $2p$ states of O represent a large contribution to the VBT of the $BiVO_4$ system, the $6s$ electrons of Bi are a large component as well.

PHOTOCATALYTIC PROPERTIES

In Figure 2, we report the wavelength dependence of the photo catalytic activity, under light irradiation from full arc up to a wavelength of 600 nm. The measurements were performed using different cut-off filters. The data acquisition time was 10 hours. Further details about the experimental set-up are described in reference [2]. The $InVO_4$ system showed not only a high photo-catalytic activity in the UV region, but also an evident activity in the visible light range. The photo catalytic activity decreased almost linearly as a function of the (increasing) wavelength from the UV range to 600 nm. This corresponds to an almost constant H_2 production rate along the whole spectral range considered and the H_2 evolution could still be observed when the wavelength of the cut-off filter was larger than 600 nm. This is a rather clear proof that the catalyst is active in a very wide wavelength range. If we compare these results with the $4d$ and $5d$ metal compounds $InNbO_4$ and $InTaO_4$, which were shown to be active in a wavelength range < 500nm, the $InVO_4$ system - which includes a $3d$ transition metal - shows a much wider light-response range, as can be theoretically expected from the differences in their related band gap and electronic structure.

DISCUSSION

There is a long-standing controversy about the V-based photo-catalysts. It has been believed so far that a system like $InVO_4$ could not reduce H^+ into H_2, since the isolated atomic d level of V (-12.55 eV) is located at a much lower value, on the same energy scale, than the analogous d states of Zr (-8.46 eV), Ta (-9.57 eV), Nb (-10.03 eV) and Ti (-11.04 eV) [11]. Indeed, a previous work on the $BiVO_4$ photo-catalyst stated that the $BiVO_4$ was not able to reduce H^+ to H_2 [12]. This suggested that this difficulty to reduce the H^+ into H_2 is a general feature of all the vanadium-oxide based systems. This statement would be correct if the location of the CBB would depend only on the positions of the energy levels of the isolated atoms. However this is not generally true. Other important factors come into play in the pinning of the CBB: the Madelung potential of the system (1), the polarization (screening) effect - which is generally small - (2) and the band broadening (3) in the extended crystal. A schematic rationalization of those effects has been given in [13].

From that schematization, it can be inferred that the dominant factor in the location of both the CBB and the VBT is the Madelung potential, which is related to the ionic field. On these grounds, it is then clear that any extrapolation from the energy spectrum of a single isolated atom suffers from a serious drawback. The whole bulk crystal environment has to be kept into account, with particular emphasis to the ionic potential at the surface, where the environment changes abruptly and where important structural modifications (e.g. relaxations and reconstructions) occur during the catalysis reaction and the related absorption of water molecules. We pushed our first principles molecular dynamics (MD) investigations [14] in this direction and work is now still in progress. The details concerning the absorption process of water molecules will be presented in a forthcoming paper [7,15]. For the ongoing discussion, the results of our *ab initio* MD simulations can be summarized as follows. We started a simulation using a periodically repeated slab of $InVO_4$ in which the exposed surface was the In rich (001). On the bottom side, the atoms were kept fixed to the bulk crystal while all the rest of the structure was allowed to relax/reconstruct according to the dynamics. One H_2O molecule was put in the vicinity of the surface and the system was allowed to evolve according to the Car-Parrinello equations of motion. We observed that the In atom has a remarkable interaction with the electron lone pairs of the O atom of the H_2O molecule and that this interaction is attractive. As a response to the approach of the water monomer, exposed In atom of the surface is displaced of about 0.5 Å from its original crystallographic position. This displacement is huge if compared to simple relaxation processes affecting e.g. the O atoms near the surface (~0.2-0.3 Å), where no interactions with water occur. Also the displacement of the V atoms, which are second neighbors not facing directly the surface in this configuration, are negligible (~0.04 Å) in comparison with In. Thus, this can be regarded as a true *strong* interaction between the catalyst surface and water. Since all these geometrical modifications are expected to affect the electronic structure and hence the energy levels distribution, we are now investigating those changes at a more detailed level. Here, we can anticipate that both the VBT and the CBB are more or less affected according to the strength of the interaction and the degree of relaxation of the system.

Another important factor, mentioned above, responsible for the location of the CBB and the VBT is the band width. A remarkable feature of the DOS of $InVO_4$ is the fact that the peaks that originate from the $3d$ electrons of V and from the $2p$ electrons of O are much sharper than the analogous ones in the $InNbO_4$ and $InTaO_4$ systems. The fact that this band is so narrow in $InVO_4$ would not allow the CBB to shift too much; on the contrary it would result almost in a *pinning* of the CBB at higher values on the energy axis with respect to a broader DOS. This feature might play a role in the experimentally evidenced ability of $InVO_4$ to reduce H^+ into H_2. A tentative explanation of the sharp DOS feature can be the following one: as mentioned above, the V-V distance (4.05 Å) in the $InVO_4$ crystal is much larger than the typical M-M distance (3.3 Å) in the $InMO_4$ systems (M = Nb, Ta). As a consequence, little overlap is expected from the $3d$ orbitals of V and small or zero overlap integral, in the electronic structure calculation jargon, means sharp peak. If this is true, then keeping a large separation between V-V might be one of the key points to realize a good catalyst. We notice, in passing, that although the atomic $3d$ level of V is located below the $3d$ level of Ti by 1.5 eV on the same energy scale, the conduction bottom level position of $InVO_4$ turns out to be almost coincident with that of the TiO_2 rutile structure. Yet, this is not so surprising and can be partially ascribed to the difference in the conduction band width of $InVO_4$ and TiO_2, as reported in reference [7,16].

A further remark that must be mentioned is the fact that despite the fact that the CBB level of $BiVO_4$ is almost identical to that of $InVO_4$, only $InVO_4$ is able to generate H_2. An exhaustive explanation is still a matter of debate, however we can argue that the difference in the electron mobility of the two conduction bands is one of the reasons that can be invoked. As shown in Table II, in $InVO_4$ the 19% of the $5s$ orbitals of In contributes to the CBB; instead, the CBB of $BiVO_4$ consists of relatively localized V_3d and O_2p states. The $5s$ wavefunction of In is a very broaden and diffused orbital that can easily extend over a distance of 3.3 Å: this is enough to make $5s$-$5s$ orbital overlap of two subsequent In atoms very large. Therefore, we might infer that the electron mobility of the CBB of $InVO_4$ is larger than that of $BiVO_4$ and this enhances the electron transfer and hinders the carrier recombination. A comparison of the VBT of $InVO_4$ and $BiVO_4$, might suggest that $InVO_4$ should display a grater catalytic ability in the evolution of O_2, since the VBT level of $InVO_4$ is located at lower values on the energy axis with respect to that of $BiVO_4$ and this should lead to a stronger oxidation power of $InVO_4$. On the other hand, experiments of O_2 yield have shown that $BiVO_4$ can act as a catalyst while such an oxygen evolution activity is absent in $InVO_4$ in the visible range. This phenomenon can be explained on the basis of the same electronic structure arguments used above. The VBT of the $InVO_4$ system consists almost of $2p$ electrons of the O atoms, which are localized. Conversely, the VBT of $BiVO_4$ has a large (18%) $6s$ component coming from Bi and this $6s$ component could be responsible for a larger carrier mobility around the top of the valence band.

CONCLUSIONS

We have computed and compared the electronic structure of $InVO_4$, $InNbO_4$, $InTaO_4$ and $BiVO_4$ in order to systematically discuss the photo-catalytic properties characterizing these different compounds. We could clarify that the conduction band bottom is characterized by a non-negligible contribution of the In_5s orbitals and a dominant d orbitals component in all the $InMO_4$ (M = V, Nb, Ta) systems. On the other hand, the valence band top of $BiVO_4$ displays Bi_6s components as well as a dominant O_2p character. By reasoning on the electronic structure, we have proposed that the photo catalytic activity might be improved by an increase in the mobility induced by the broad s orbital component as well as by tuning the location of both the CBB and the VBT. This latter issue rely, of course, on the accurate choice of the compound and its crystallographic structure. From first principles molecular dynamics simulation, we have seen that the electronic structure modifications at the active surface play an important role. These arise from relaxations and electrostatic interactions with incoming water molecules that induce dramatic changes at the catalytic sites on the surface. All these processes alter significantly the bulk atomic and electronic structure, therefore they must be kept into account for a full understanding of the photo-catalytic processes.

ACKNOWLEDGEMENTS

We are grateful to Dr. Ferdi Aryasetiawan for insightful discussions and precious suggestions in the LMTO calculations.

REFERENCES

1. Z. Zou, J. Ye, K. Sayama and H. Arakawa, *Nature* **414**, 625 (2001).
2. J. Ye, Z. Zou, M. Oshikiri, A. Matsusita, M. Shimoda, M. Imai and T. Shishido, *Chem. Phys. Lett.* **356**, 221 (2002).
3. A. Kudo, K. Omori and H. Kato, *J. Am. Chem. Soc.* **121**, 11459 (1999).
4. Z. Zou, J. Ye and H. Arakawa *Chem. Phys. Lett.* **332**, 271 (2000).
5. P. M. Touboul and P. Toledano, *Acta Cryst.* **B36**, 240 (1980).
6. P. J. Bridge and M. W. Pryce, *Mineralogical Magazine* **39**, 847 (1974)
7. M. Oshikiri, M. Boero, J. Ye, Z. Zou and G. Kido, *J. Chem. Phys.* **117**, (2002) in press.
8. M. I. Kay, B. C. Frazer and I. Almodovar, *J. Chem. Phys.* **40**, 504 (1964)
9. O. K. Andersen, *Phys. Rev.* **B 12**, 3060 (1975); O. K. Andersen and O. Jepsen: *Phys. Rev. Lett.* **53**, 2571 (1984).
10. L. Hedin and S. Lundqvist, *Solid State Physics* **23**, 1 (1969); M. S. Hybertsen and S. G. Louie, *Phys. Rev.* **B 34**, 5390 (1986); F. Aryasetiawan, *Phys. Rev.* **B 46**, 13051 (1992); M. Oshikiri and F. Aryasetiawan, *Phys. Rev.* **B 60**, 10754 (1999).
11. W. A. Harrison, "Electronic Structure and the Properties of Solids – The Physics of the Chemical Bond", (W. H. Freeman and Company, 1980).
12. A. Kudo, K. Ueda, H. Kato and I. Mikami, *Catalysis Lett.* **53**, 229 (1998).
13. P. A. Cox, "The Electronic Structure and Chemistry of Solids", (Oxford University Press, 1987).
14. R. Car and M. Parrinello, *Phys. Rev. Lett.* **55**, 2471 (1985); CPMD code by J. Hutter et al., Max-Planck-Institut für Festkörperforschung and IBM Zurich Research Laboratory, 1995-1999.
15. to be submitted to the 26 th International Conference on the Physics of Semiconductors (2002).
16. M. Oshikiri and J. Ye, Extended Abstracts 12p-YA-16 (The 62 th Autumn Meeting, 2001); The Japan Society of Applied Physics Society.

Mat. Res. Soc. Symp. Proc. Vol. 730 © 2002 Materials Research Society   V5.9

Texture Evolution in Zr Grain-refined by Equal Channel Angular Pressing

S.H. Yu[1], H.S. Ryoo[1], D.H. Shin[2] and S.K. Hwang[1,3]
[1]School of Materials Science and Engineering, Inha University,
Incheon, Korea, 402-751
[2]Department of Metallurgy and Materials Science, Hanyang University,
Ansan, Gyunggi-Do, Korea, 425-791
[3]Jointly appointed by the Center for Advanced Aerospace Materials,
Pohang University of Science and Technology, Pohang, Korea

ABSTRACT

To explore a possibility of obtaining an ultra fine grain size in hcp materials, equal channel angular pressing (ECAP) was applied to commercially pure Zr with the initial grain size of 20 μm. To ensure the crack-free specimen during the severe deformation it was necessary to adopt different die designs at the two processing temperatures, room temperature and 350°C. Sub-micrometer scale grains, in the order of 200 nm, were obtained by employing the die design of 90°/20° and the deformation temperature of 350°C. With the proper choice of the processing parameters, the refined grains were surrounded by the high angle boundaries. During the severe plastic deformation, the crystal texture underwent a significant change, from a fiber texture to that of a strong (0002) component mixed with a medium $\{10\bar{1}0\}$ components, suggesting a dual <a> slip mode of deformation on the plane of the maximum shear stress.

INTRODUCTION

Zirconium, having a low neutron cross section and a high resistance to corrosion, is an important metal in nuclear energy applications. Many components in the nuclear reactor core are made of Zr, fuel cladding and pressure tubes being examples of the primary utilization of Zr-base alloys. Current Zr-base alloys have been designed on the basis of alloying additions such as Nb and Sn to enhance the structural properties related to nuclear energy generation such as the yield strength and the corrosion resistance. With the recent advances in the ultra-fine grain technologies of bulk metals, however, it is of interest to investigate the possibility of applying this non-conventional strengthening route to Zr, which might prove to be attractive in terms of the long-term economic advantage since the components may become smaller in dimension. In this respect, equal channel angular pressing (ECAP) is considered to be a potential means of achieving improved mechanical as well as the electrochemical properties in this metal. Despite its wide success in metals of cubic crystal structure [1-5], ECAP application in materials of hcp system is relatively scanty except for Mg and Ti. So far there is no report of the related research attempt in Zr. In the present study, we address the possibility of grain refining Zr through a severe plastic deformation process. Along with the grain refinement, the issue of texture has been also addressed in the context of the deformation mechanism since it is another important attribute in the performance of the nuclear core components.

EXPERIMENTAL PROCEDURE

Commercially pure zirconium (Zr702) containing 4.5 wt.% Hf, 0.20 wt.% (Fe+Cr), 0.005 wt.% H, 0.005 wt.% N, 0.05 wt.% C and 0.16 wt.% O was used as the starting material. Two types of specimen shapes were used, a cylindrical type of 10 mm in diameter and 70 mm long and a square bar type of 10 mm × 10 mm × 70 mm. The initial materials were in the as-received (hot-extruded) state without additional heat treatment. For the cylindrical sample, hereafter named as Zr(C), the ECAP die was designed to yield a shear strain of approximately 1.83 in a single pass: the inner contact angle (ϕ) and the arc of curvature (ψ) at the outer point of contact between channels of the die were 90° and 20°, respectively. For the square bar type sample, hereafter name as Zr(B), another ECAP die was designed with the angles of ϕ=135° and ψ=45°. The ECAP process temperatures for the two types of specimen, Zr(C) and Zr(B), were 350°C and room temperature, respectively. With the die design of 90°/20°, it was impossible to produce defect-free specimens at room temperature. ECAP was conducted for both types of dies at a constant pressing speed of 2mm/s, using MoS_2 to minimize the friction. Schematics of the die design and the work schedules are shown in figure 1. Transmission electron microscopy (TEM) samples were sliced normal to the longitudinal axis, i.e., the X-axis in figure 1. TEM images were taken with Phillips CM200, operating at an accelerating voltage of 200 kV. For the texture analysis, X-Ray diffraction (XRD) specimens were also cut normal to the X-axis. Using Seifert XDAL3000 with Cu target operating at the accelerating voltage and current of 40 kV and 30 mA, respectively, the pole figures of (0001), ($10\bar{1}0$), ($11\bar{2}0$) and ($10\bar{1}1$) planes were obtained.

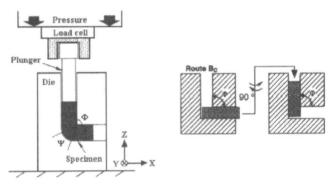

Figure 1. Definition of the plane orientations and the rotation scheme of ECAP route. In this figure the die shape is for 90°/20° that was used for cylindrical shape Zr702 processed at 350°C. Other die design with 135°/45° configuration was used for processing the square bar type Zr702 specimen at room temperature.

RESULTS AND DISCUSSION

Zr(B) deformed at room temperature had clean surfaces free of cracks. However, in the surface of the Zr(B) specimen processed at room temperature, a wavy feature indicative of localized slips appeared after ECA deformation, in contrast to the smooth surface of Zr(C) processed at 350°C.

In terms of the microstructure, there were significant differences between the two types of specimen. A significant grain refinement occurred in Zr(C), which was deformed at 350°C using the die design with 90°/20°. In TEM analysis, the average grain size of Zr(C) specimens was reduced to approximately 270 nm in size by a single pass ECA deformation. This is compared to the average grain size of 20 μm in the starting material. The grain boundaries in this state, however, showed an inclination toward the low angles in misorientation, hinting a considerable volume fraction of subgrains. Increasing the number of passes via the route B_C resulted in a further grain refinement, down to 200 nm. This scale of grain refinement must be regarded as genuine because the selected area diffraction pattern in TEM, figure 2(a), showed essentially a ring pattern, which verifies the high angle random grain boundaries among the refined grains.

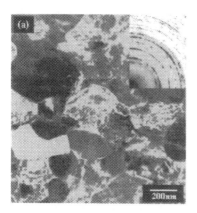 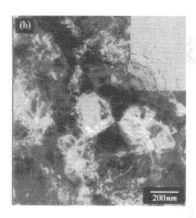

Figure 2. TEM bright field micrographs with the insets of selected area diffraction patterns of ECA deformed zirconium: (a) Zr702 after four passes via route B_C using 90°/20° die at 350°C and (b) sponge type zirconium after eight passes via route B_C using 135°/45° die at room temperature reported in an earlier work [6].

In contrast to the specimen Zr(C) deformed in the 90°/20° die, the specimen Zr(B) deformed in the 135°/45° die did not show appreciable grain refinement in four passes. Unlike the former, the latter specimen neither showed subgrain-like microstructural refinement in the single pass nor showed noticeable grain refinement with the high angle boundaries in four passes via the route B_C. In this respect, it is interesting to compare the present result with the result obtained earlier in the high purity sponge type Zr. According to Choi et al. [6], the average grain size was reduced from 200 μm to 400 nm in four passes via the route B_C in the high purity sponge type Zr using the same type of die and the deformation temperature - room temperature. Furthermore, increasing the number of passes from four to eight resulted in an additional grain refinement from 400 nm to 200 nm with mostly high angle grain boundaries - see figure 2(b).

Difference in the die design may be considered as a possible source of microstructural variation. Using the Iwahashi's formula [7], the amount of shear strain in a single pass of the two modes of pressing were approximately 1.83 and 0.79 for the 90°/20° design and 135°/45° design, respectively. Therefore higher degree of plastic deformation is expected in the former scheme whereas more passes are required for the latter scheme to accumulate the threshold amount of plastic deformation. Despite the penalty, the 135°/45° die design is still of value in case the grain refinement temperature is limited to room temperature. Besides the die design, the impurity level of the specimens may also be essential in accounting for the difference in the extent of grain refinement, details of which are the subject of the future research.

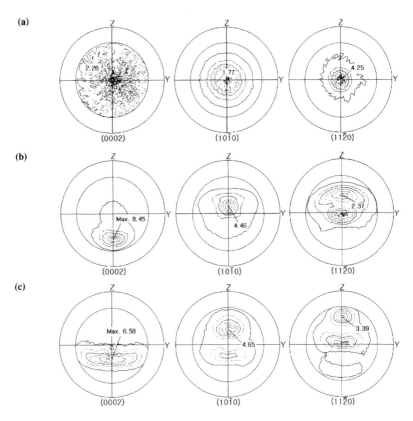

Figure 3. Pole figures of Zr702 showing the progress of the texture evolution during ECA deformation: (a) prior to ECAP and (b) after single ECA pass in 90°/20° die and (c) after single ECA pass in 135°/45° die.

In addition to the grain refining effect, ECAP produced a remarkable change in crystal texture. The texture variation in two different specimens is shown in figure 3. In the as-received state, Zr702 showed an <a> fiber texture that can be described as a strong alignment of the {11$\bar{2}$0} poles with the X-axis and the random distribution of other poles as shown in figure 3(a) regardless of the specimens shape. Due to the symmetry of the crystal, the (0002) poles are supposed to be distributed evenly along the periphery of the Y-Z plane although this feature is missing in figure 3(a) because the pole distributions up to 70° only are shown in this figure.

During the ECA deformation the crystal texture underwent a significant change in both specimens. The <a> fiber texture in the as-received Zr702 turned into a strong basal/prism texture as shown in figure 3(b) and (c). In Zr(C), which was deformed using the 90°/20° die, the sudden change of the initial texture to the basal/prism texture during the single pass of ECAP proves the severity of deformation. As shown in figure 3(b), the main feature of the basal/prism texture in deformed specimen is the intense concentration of the two poles along a certain orientation within the specimen. The specific crystal orientations of the two poles, <c> (basal) and <p> (prism), are about 55° and 15-20° rotations, respectively, with respect to the Y-axis. In case of the <c> pole, the intensity reached as high as 8.5. Therefore this phenomenon can be interpreted as the crystal rotation to accommodate the severe shear deformation during ECAP.

In case of Zr(B), which was deformed using the 135°/45° die, the crystal texture in the ECA deformed specimen became much like that in Zr(C) qualitatively but not quantitatively. As shown in figure 3(c), a rather strong concentration of the <c> poles, reaching an intensity of 6.58, was found at 30° from the X-direction on the X-Z plane. Concurrently the <p> poles appeared at 30-35° on the same plane. The deviation of the locations of the poles from those in specimen Zr(C) is due to the different amount of the effective strain during ECAP because of the difference in the die design. However it is further noted that the poles, particularly the (0002) poles, became more dispersed than those in specimen Zr(C). The common feature of the two specimens, the location of the strong basal poles and the prism poles on the maximum shear plane, indicates that the major deformation mode in both specimens was the <a> slip.

Based upon the results of the texture analysis, the mechanism of crystal rotation during ECAP of Zr was sought. In case of Zr(C), either basal slip on (0002) or prismatic slip on $(10\bar{1}0)$ was activated with the common slip direction of <a>, i.e., $<11\bar{2}0>$. Assuming that the major mode of deformation during ECAP is pure shear, the planes of maximum resolved shear stress are easily identified. Basically two types of the <a> slips, basal and prismatic, predominate the deformation. The schematics of the shear planes created by ECAP and relevant rotations of crystals in the pole figure are explained in terms of the <a> shear in figure 4.

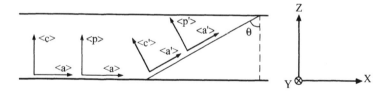

Figure 4. Schematic of the shear planes created by ECAP and pertinent crystals rotation in the pole figure of Zr-C: (a) the basal <a> slip and the prismatic <a> slip acting on one shear plane (on the Z-plane, denoted as <a>, <p> and <c>, with <p> meaning a $<10\bar{1}0>$ prism pole) and on another shear plane (on the inclined plane, denoted as <a'>, <p'> and <c'>).

Although the <a> slip was the major active mechanism of deformation, twinning may also have contributed to deformation in specimen Zr(B). Deformation twins were observed in Zr(B) but not in Zr(C) nor in the sponge type Zr that underwent multiple passes of ECAP. It is interesting to note that the density of deformation twins in Zr during cold rolling reaches maximum when the strain per each pass is comparatively low, about 20% [8]. Therefore it may be suggested that certain low level plastic strain activates twinning in Zr while a severe or accumulated strains do not. It is unclear, however, how the purity, deformation temperature and the strain rate affect twinning.

CONCLUSIONS

Evolution of the microstructure and texture of commercially pure Zr were studied with the ECAP as a means of grain refinement through the severe plastic deformation. Starting from the initial size of 20 μm, it was possible to reduce the average grain size down to a level of 200 nm by employing a die design of 90°/20° and the deformation temperature of 350°C. The efficiency of grain refinement sensitively depended on the die design and the deformation temperature mainly because of the amount of the effective strain to be attained in each pass. During the severe deformation, the crystal texture radically changed from a fiber texture to the basal plus prismatic texture. The major deformation mechanism, therefore, was concluded to be the <a> slip, with a minor possibility of twinning in the case of the less severe die design.

ACKNOWLEDGEMENTS

The present work was performed under the auspices of the Korean Ministry of Science and Technology through the '2001 National Research Laboratory Program'.

REFERENCES

1. S.L. Semiatin, P.B. Berbon and T.G. Langdon, Scripta Mater. **44**, 135 (2001).
2. K. Neishi, Z. Horita and T.G. Langdon, Mater. Sci. Eng. **A325**, 54 (2002).
3. Y. Fukuda, K. Oh-ishi, Z. Horita and T.G. Langdon, Acta Mater. **50**, 1359 (2002).
4. K-T Park, Y-S Kim and D.H. Shin, Metall. Trans. **A32**, 2373 (2001).
5. Z. Horita, T. Fujinami, M. Nemoto and T.G. Langdon, J. of Mater. Proc. Tech. **117**, 288 (2001).
6. W.S. Choi, H.S. Ryoo, S.K. Hwang, M.H. Kim, S.I. Kwun and S.W. Chae, Metall. Trans. **A33**, 973 (2002).
7. Y. Iwahashi, J. Wang, Z. Horita, M. Nemoto and T.G. Langdon, Scripta Mater. **35**, 143 (1996).
8. D.L. Douglass in *The Metallurgy of Zirconium*, edited by Zh.I. Turkoxv and D. Twersky, (IAEA, Austria) p. 45.

Mat. Res. Soc. Symp. Proc. Vol. 730 © 2002 Materials Research Society

Chemically deposited antimony selenide thin films

Y. Rodríguez-Lazcano, Y. Peña, M. T. S. Nair, and P. K. Nair
Centro de Investigación en Energía, Universidad Nacional Autónoma de México
Temixco, Morelos 62580, MÉXICO. mtsn@cie.unam.mx

ABSTRACT

Chemical bath deposition of thin films of antimony selenide from aqueous solutions containing complexes of antimony with citrate, tartrate and thiosulfate as ligands and sodium selenosulfate as source of selenide is reported. The films obtained appear amorphous in the as-prepared form and become crystalline upon annealing at 300°C. The X-ray Diffraction (XRD) patterns of the annealed films show peaks attributable to Sb_2Se_3 and Sb_2O_3. Electron microprobe analyses have shown that the atomic ratio of Se/Sb is less than 1.5 in these films. The films are photoconductive and exhibit a high resistivity in the dark. Both direct (1.4 eV) and indirect (1.3-1.5 eV) band gaps are observed for the films.

INTRODUCTION

Antimony selenide (Sb_2Se_3) with an optical band gap of 1.11 eV [1] is considered appropriate for application as the absorber component in polycrystalline thin film solar cells. Methods reported for obtaining thin films containing the material include vacuum evaporation [2], spray pyrolysis [3], electrodeposition [4, 5] and chemical bath deposition [6,7]. Thin films obtained by chemical bath deposition are reported as amorphous, while those obtained by the other methods are crystalline. In some cases crystallinity is achieved only after the films have been annealed at temperatures up to 200°C.

Chemical bath deposition has the advantage of being a low capital-intensive technology for obtaining thin films of metal chalcogenides over surfaces of large areas and of any shape. Hence the technique is widely accepted as suitable for depositing thin films for solar energy related applications [8]. The use of chemically deposited antimony selenide thin films as photoactive material in photoelectrochemical cells has been demonstrated [7]. In the present work, we deposited antimony selenide thin films from different chemical baths with a view to investigate the possibility of improving the crystallinity of the films by post deposition treatments and to analyze their electrical and optical properties.

EXPERIMENTAL DETAILS

Deposition of thin films: Clean microscope glass slides were used as substrates for the deposition of antimony selenide thin films from three different baths prepared by using antimony trichloride or potassium antimony tartrate as described below:
Bath # 1: One gram of antimony trichloride ($SbCl_3$ -Fermont) was transferred to a 100 ml beaker. To this was added 37 ml of 1M solution of sodium citrate (Baker Analyzed Reagent), with stirring. Upon the addition of citrate, a white precipitate is formed initially, which dissolves

in excess of the reagent as the addition continues. This was followed by the sequential addition of 20 ml of 30% ammonia(aq) (Fermont), 24 ml of 0.4 M sodium selenosulfate (prepared in the laboratory) and the rest deionized water to take the volume up to 100 ml. The mixture was stirred throughout the addition and was devoid of any precipitate.

Bath # 2: Twenty-five ml of 0.1 M potassium antimony tartrate (Baker Analyzed Reagent) was taken in a 100 ml beaker. To this was added with stirring 2 ml of approx. 3.7 M triethanolamine (Baker Analyzed Reagent) followed by 20 ml of 30 % ammonia(aq), 10 ml of 0.4 M sodium selenosulfate and the rest de-ionized water to make up to 100 ml.

Bath # 3: Five hundred mg of antimony trichloride was dissolved in 2.5 ml of acetone in a 100 ml beaker. This was followed by the addition of 20 ml of 1 M sodium citrate, 15 ml of 30% aqueous ammonia, 10 ml of 1 M sodium thiosulfate (Baker Analyzed Reagent), 20 ml of 0.1 M sodium selenosulfate and the rest deionized water to complete the volume to 100 ml.

In all the cases the starting solution mixture was clear. Glass substrates were introduced vertically in the bath, supported against the wall of the beaker. The depositions were carried out at 27°C in the case of baths 1 and 2 and at 10°C in the case of bath 3. A Polyscience digital temperature controller circulation bath was used. The substrates coated with the films were taken out of the baths at different intervals of time ranging from 1 to 6 h. These were washed with distilled water and dried by blowing hot air. Both sides of the substrates were coated with specularly reflecting brown films, characteristic color of antimony selenide. The thin film deposited on the side near the wall of the beaker was kept intact in all the cases for characterization.

Thickness of the films was measured in an Alfa Step 100 unit (Tencor Inc., CA, USA). The thickness of the films varied in the ranges of 0.2 μm (1 h) - 0.7 μm (5 h), 0.1 μm (2 h) - 0.4 μm (4 h) and 0.6 μm (4 h) - 0.8 μm (6 h), respectively, from baths 1, 2 and 3, respectively, for the durations of depositions given in parentheses. The thin films were annealed in a vacuum oven (T-M High Vacuum Products) in a nitrogen atmosphere of 100 milliTorr at temperatures ranging from 150-350°C for 1 h each.

Characterization: X-ray diffraction (XRD) patterns of the films were recorded on a Rigaku x-ray diffractometer using Cu-K$_\alpha$ radiation. The morphology was studied by scanning electron microscopy (SEM) and the composition by electron probe microanalysis (EPMA). Optical transmittance (T%) and reflectance (R%) spectra of the films with air and a front aluminized mirror as references, respectively, were recorded on a Shimadzu UV-3101PC UV-VIS-NIR scanning spectrophotometer. For the electrical measurements, a pair of coplanar silver paint electrodes was printed on the films. The current against time data were recorded with the samples in the dark and under illumination with 2 kW m^{-2} tungsten-halogen light on a computerized system using a Keithley 619 electrometer and a Keithley 230 programmable voltage source. For the initial 20 s the sample was in the dark, followed by 20 s under illumination and finally for 20 s in the dark.

RESULTS AND DISCUSSION

XRD patterns of the films from the three baths did not show any peak in the as-prepared form. Well-defined peaks are seen in XRD patterns of the films annealed at 300°C, as shown in Fig. 1 for the case of films obtained from bath 3. The pattern shows peaks matching those of Sb$_2$Se$_3$ (JCPDS 15-0861) and Sb$_2$O$_3$ (JCPDS 43-1071), with the peaks of the latter dominating the pattern. In the case of films obtained from baths 1 and 2, the peaks due to Sb_2Se_3 component are still weaker. Thus, the films obtained from the three baths employed here contain Sb_2O_3 along with Sb_2Se_3 that can be crystallized by the annealing. Such post deposition annealing leading to crystallization has been reported for thin films of Sb_2Se_3 obtained by thermal evaporation [2].

The presence of Sb_2O_3 component in the films is a consequence of insoluble hydroxide formation in non-acidic aqueous solutions of group V metal ions: $Sb^{3+} + 3H_2O \rightarrow Sb(OH)_3 + 3H^+$; $2Sb(OH)_3 \rightarrow Sb_2O_3 + 3H_2O$. In the present case, ammonia(aq) in the bath will further facilitate the hydroxide formation. We have observed a similar effect of co-deposition of oxide in the case of bismuth selenide films formed from solutions containing bismuth nitrate, triethanolamine and dimethylselenourea at temperatures above 40°C or selenosulfate and ammonia at room temperature.

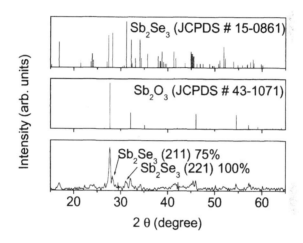

Figure 1. XRD pattern of a sample of antimony selenide annealed at 300°C

Composition analyses by electron microprobe (EPMA) on the samples have shown that the atomic ratio of Se/Sb is less than 1.5. The relative excess of Sb corroborates the XRD data in Fig. 1 showing the presence of the oxide phase.

Figure 2 shows the optical transmittance (T%) and reflectance (R%) spectra of the films deposited for 4 h in the three different baths and subsequently annealed in nitrogen for 1 h each at 300°C. The transmittance data are corrected for the loss due to reflectance:
Tcorr % = 100 T% / (100 − R %). Figure 3 gives the corrected transmittance for the films obtained from bath 3 annealed at different temperatures. From the corrected transmittance curves, values of absorption coefficient α were computed using the formula, $\alpha = (1/d)\ln(100/Tcorr)$ for different wavelengths. Plots of α^2 and $\alpha^{1/2}$ against photon energy (E= hv) are used to obtain the optical band gaps (E_g) in the films for direct and indirect transitions, respectively. Extrapolating the straight line part in such plots given in Figure 4 for the films in

the present study gives an indirect band gap of 1.3 eV and 1.5 eV for samples from baths 1 and 2, respectively, and a direct band gap of 1.4 eV for samples from bath 3. Both direct and indirect band gaps are reported for antimony selenide in bulk [1] and thin film forms [2,4].

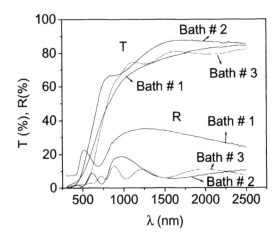

Figure 2. Transmittance (T%) and reflectance (R%) spectra of the films from the different baths, after annealing at 300°C

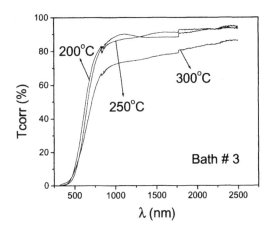

Figure 3. Corrected transmittance curves (Tcorr %) for the films obtained from bath 3 after annealing at different temperatures.

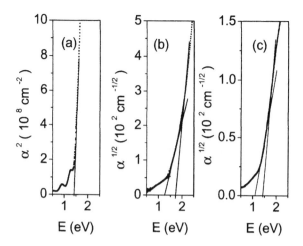

Figure 4. Plots of squares and square roots of absorption coefficients against photon energy, showing direct and indirect band gaps in the films annealed at 300°C: (a) bath 3, $E_g = 1.4$ eV, direct; (b) bath 2, Eg = 1.5 eV, indirect; and (c) bath 1, Eg = 1.3 eV, indirect.

Figure 5 shows the photocurrent response curves of the thin films annealed at 300°C during 1 h. The films show dark conductivities of 2.0×10^{-8}, 2.6×10^{-9} and 4.6×10^{-10} Ω^{-1} cm^{-1} for the samples obtained from baths 1, 2 and 3, respectively. In all three cases, the films are photoconductive.

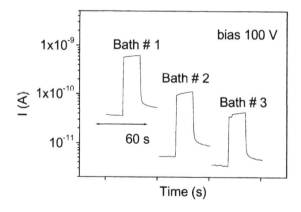

Figure 5. Photocurrent response curves for the samples annealed at 300°C

CONCLUSIONS

Thin films of antimony selenide obtained from three different chemical baths appear amorphous in the as-prepared form in XRD patterns. Annealing the films at 300°C in vacuum oven in 100 mTorr nitrogen pressure introduces crystallinity. The annealed films contain crystalline Sb_2Se_3 and Sb_2O_3 phases. The oxide is incorporated in the film from the baths as antimony ions have a strong tendency to form insoluble hydroxide leading to oxide in aqueous solutions that are not highly acidic. The annealed films show both direct and indirect band gaps in the range of 1.3 eV – 1.5 eV depending on the composition of the bath from which they are formed. The electrical conductivity is in the range of 10^{-10} to $10^{-8} \Omega^{-1} cm^{-1}$.

ACKNOWLEDGMENTS

The authors are grateful to Leticia Baños and to José Guzmán, of IIM, UNAM for recording the XRD patterns and for the SEM and EPMA analysis, respectively. We acknowledge the financial support received from DGAPA-UNAM (IN105700), CONACyT (34368-E) and DGEP-UNAM.

REFERENCES

1. O. Madelung (editor) in *Semiconductors: Other than group IV elements and III-V compounds* (Springer-Verlag Berlin Heidelberg, 1992) p.50.
2. H. T. El-Shair, A. M. Ibrahim, E. Abd El-Wahabb, M. A. Afity and F.Abd El-Salam, *Vacuum*, **42**, 911 (1991).
3. K. Y. Rajpure and C. H. Bhosale, *Materials Chemistry and Physics*, **62**, 169 (2000).
4. A. P. Torane, K. Y. Rajpure and C. H. Bhosele, *Materials Chemistry and Physics*, **61**, 219 (2000).
5. A. M. Fernández and M. G. Merino, *Thin Solid Films*, **366**, 202 (2000).
6. P. Pramanik and R. N. Bhattacharya, *J. Solid State Chem.*, **44**, 425 (1982).
7. R. N. Bhattacharya and P. Pramanik, *Solar Energy Materials*, **6**, 317 (1982).
8. P. K. Nair, M. T. S. Nair, V. M. García, O. L. Arenas, Y. Peña, A. Castillo, I. T. Ayala, O. GomezDaza, A. Sánchez, J. Campos, H. Hu, R. Suárez and M. E. Rincón, *Solar Energy Materials and Solar Cells*, **52**, 313 (1998).

Mat. Res. Soc. Symp. Proc. Vol. 730 © 2002 Materials Research Society   V5.12

Electrical and Optical characteristics of ZnO:Al Thin Films

M.J.Keum[1], J.S.Yang[1], I.H.Son[2], S.K.Shin[3], H.W.Choi[1], W.S.Lee[4], M.K.Choi[4], S.N.Chu[4], K.H.Kim[1]

[1]School of Electrical & Electronic Eng., Kyungwon Univ., Kyunggi-do 461-701, Korea
[2]Dept. of Electrical Eng., Shinsung College, Chungnam 343-861, Korea
[3]Dept. of Electrical Eng., Donghae Univ., Kangwon-do 240-713, Korea
[4]Dept. of Electric Control System, Kyungwon College, Kyunggi-do 461-701, Korea

ABSTRACT

ZnO with hexagonal wurzite structure is a wide band gap n-type semiconductor. ZnO films can be prepared to obtain high transparency in the visible range, low resistivity, chemical stability and stability in hydrogen plasma including many foreign materials such as Al, In. In this work, we prepared ZnO:Al thin film by Facing Targets Sputtering system with Zn metal target and ZnO:Al(Al$_2$O$_3$ 2wt%, 4wt%) ceramic target at total working gas pressure 1mTorr, substrate temperature R.T.. We evaluated the crystallographic, electrical and optical characteristics of the ZnO:Al films.

INTRODUCTION

The increasing use of transparent conductive electrodes for flat panel display devices and solar cells has promoted the development of inexpensive materials such as zinc oxide. The electrical, optical and crystallographic properties of ZnO are strongly affected by impurities, which may be added in controlled amounts. ZnO films can be prepared to obtain high transparency in the visible range, low resistivity, chemical stability and stability in a hydrogen plasma containing ions such as Al, In, Si, F and Ga [1-3].

Conductive transparent ZnO:Al films with low resistivity for flat panel displays can be prepared by sputtering methods [4,5]. Especially, it was reported [6] that the sputtering conditions such as substrate temperature, deposition rate, location of the substrate and working pressure are influence the crystallographic and electrical properties of sputtered ZnO thin films. In this study, we prepared ZnO:Al transparent conductive thin films at low working gas pressure (1mTorr) and room temperature using a Facing Targets Sputtering System [7].

EXPERIMENTAL DETAILS

The Facing Targets Sputtering System that has two targets. One target was a Zn metal target (upper), and the other was a ZnO:Al ceramic target (below) doped with 2wt% or 4wt% Al_2O_3, respectively.

Figure 1 shown the schematic diagram of Facing Targets Sputtering apparatus used in this study and the sputtering conditions for preparing the ZnO:Al thin films are given in Table I.

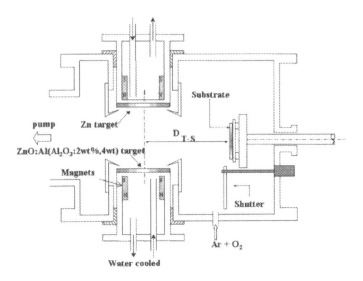

Figure 1. Schematic diagram of FTS apparatus

The oxygen gas flow rate was controlled from 0.1 to 0.3 because the transition region of the crystal structure and electric characteristic of ZnO:Al thin films were shown at oxygen gas flow rate 0.2-0.3 [8].

The crystallographic and electric characteristic was measured by x-ray diffraction (Rigaku) using CuKα radiation and 4-point probe, respectively. The optical transmittance of the ZnO:Al films were measured by UV/Visible spectrometer (Hewlett Packard).

Table I Sputtering Conditions

Deposition parameter	Conditions
Targets	Zn(4N) ZnO:Al(Al$_2$O$_3$: 2wt%, 4wt%)
Substrate	Glass slide
Target-Target Distance	100mm
Targets-Substrate Distance	100mm
Sputtering Gas	Ar, O$_2$
Base Pressure	2×10^{-6} Torr
Discharge Pressure	1mTorr
Substrate Temperature	R.T.
Sputtering Current	0.6 [A]
Oxygen Gas Flow Rate	$\dfrac{O_2[sccm]}{O_2[sccm] + Ar[sccm]}$

DISCUSSION

Figures 2 and 3 show the XRD 2θ peak of ZnO:Al (Al$_2$O$_3$:2wt%, 4wt%) thin films with film thickness 700[nm] as a function of oxygen gas flow rate. The ZnO:Al thin films structure was changed by increasing of oxygen gas flow rate, and we noticed that the XRD 2θ peak of ZnO:Al thin film prepared by Zn-ZnO:Al (Al$_2$O$_3$:2wt%) targets were highly than Zn-ZnO:Al (Al$_2$O$_3$:4wt%) targets.

It is considered that the crystallographic of ZnO:Al thin film was influenced with Al concentration in the film. The ZnO:Al thin films with a good crystal structure were prepared at oxygen gas flow rate 0.2 (Zn-ZnO:Al(Al$_2$O$_3$:2wt%)), 0.26 (Zn-ZnO:Al(Al$_2$O$_3$:4wt%)), respectively. Also, XRD 2θ peak of ZnO:Al thin film prepared by Zn-ZnO:Al (Al$_2$O$_3$:4wt%) targets show the (110) plane at oxygen gas flow rate 0.1, 0.3, respectively. But, XRD 2θ peak of ZnO:Al thin film prepared by Zn-ZnO:Al (Al$_2$O$_3$:2wt%) was changed from (002) plane to (110) plane at oxygen gas flow rate 0.3.

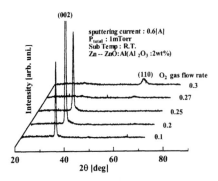

Figure 2. XRD patterns of ZnO:Al(Al$_2$O$_3$:2wt%) thin films as a function of oxygen gas flow rate

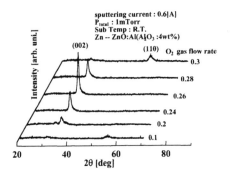

Figure 3. XRD patterns of ZnO:Al(Al$_2$O$_3$:4wt%) thin films as a function of oxygen gas flow rate.

Figure 4 show the resistivity of ZnO:Al thin film prepared by Zn-ZnO:Al(Al$_2$O$_3$:2wt%, 4wt%) targets. The resistivity of ZnO:Al thin film was low value(< 10^{-2} Ω-cm) at oxygen gas flow rate under 0.3. Especially, the resistivity of ZnO:Al thin film prepared by Zn-ZnO:Al(Al$_2$O$_3$:2wt%,) targets was lowest value(< 10^{-4} Ω-cm) at oxygen gas flow rate 0.2. But the resistivity of ZnO:Al thin film was dramatically increased at oxygen gas flow rate 0.3. With the increase of the oxygen gas flow rate (> 0.3), the resistivity of ZnO:Al thin film increased because the oxygen voids in the films were substituted for oxygen atoms and the additional oxygen atoms in the films effectively function as carrier traps [9]. Figure 5 shows the optical transmittance of ZnO:Al thin films prepared Zn-ZnO:Al(Al$_2$O$_3$:2wt%, 4wt%) targets. Optical transmittance of all films was shown over 80%. Especially, optical transmittances of ZnO:Al thin films prepared by Zn-ZnO:Al (Al$_2$O$_3$: 4wt%) target were shown close to 90% in visible range.

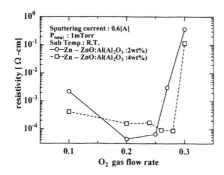

Figure 4. XRD patterns of ZnO:Al(Al$_2$O$_3$:4wt%) thin films as a function of oxygen gas flow rate.

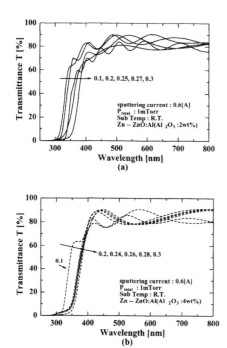

Figure 5. Variation of the transmittance on Oxygen gas flow rate

((a): Zn-ZnO:Al(Al$_2$O$_3$ 2wt%)target, (b): Zn-ZnO:Al (Al$_2$O$_3$ 4wt%) target)

XRD patterns and electric characteristic of ZnO:Al thin film was changed with the increasing of oxygen gas flow rate and Al concentration in the target. But the optical transmittance was shown similar with the increasing of oxygen gas flow rate and Al concentration in the target.

CONCLUSION

ZnO:Al transparent conductive thin films were prepared on slide glass at room temperature, sputtering current 0.6 [A], total working gas pressure 1mTorr by Facing Targets Sputtering System that has Zn-ZnO:Al (Al$_2$O$_3$: 2wt%,4wt%) targets, and investigated electrical, crystallographic and optical properties with oxygen gas flow rate.

In above experimental results, we obtained the ZnO:Al transparent conductive thin film with low reisistivity (< 10^{-4} Ω-cm) and the optical transmittance above 80% at oxygen gas flow rate 0.2 and using Zn- ZnO:Al (Al$_2$O$_3$: 2wt%) targets.

In case of, the optical transmittance of ZnO:Al thin films prepared by Zn-ZnO:Al (Al$_2$O$_3$: 4wt%) targets was highly (close to 90%) than Zn-ZnO:Al (Al$_2$O$_3$: 2wt%) targets. But electrical and crystallographic properties were shown low grade than that.

Therefore, in these results, we noticed that the ZnO:Al transmittance conductive thin film with a good electrical and optical properties can prepared at room temperature by Facing Targets Sputtering System.

REFERENCES

1. Y. Igasaki and H. Saito, J.Apply. Phys., 70, 3613 (1991)
2. T. Minami, H. Nanto, and S.Takata, Thin Solid Films, 124, pp.43-47 (1985)
3. A. Banerjce, J. Yang and S. Guha, Mater. Res. Soc. Symp. Proc., 467, 711 (1997)
4. T.Minami, H.Nanto, and S.Takata, Appl.Phys.Lett. 41,954 (1982)
5. K.Ito and T.Nakaxawa, Jpn.J.Appl.Phys.22, L245 (1983)
6. S. Maniv and A. Zangvil, "Controlled texture of reactively RF sputtered ZnO thin film", J.Apply. Phys., 49 (1978) 2787
7. K.H.Kim, S.H.Kong, M.J.Keum, I.H.Son, M.Naoe, S.Nakagawa, "Thin Film Properties by Facing Targets Sputtering system", Apply. Sur. Sci., Vol. 169, No.170, pp.409-413 (2001)
8. J.S.Yang,M.J.Keum, K.H.Kim, presented at the FES2001, Nagoya, Japan (2001)(unpublished)
9. Takashi Tsuji and Mitsuji Hirohashi, Apply. Sur. Sci., 157 (2000) 47

Mat. Res. Soc. Symp. Proc. Vol. 730 © 2002 Materials Research Society   V5.14

Absorber Films of Ag$_2$S and AgBiS$_2$ prepared by Chemical Bath Deposition

A. Nuñez Rodriguez, M.T.S. Nair and P.K. Nair
Centro de Investigación en Energía, Universidad Nacional Autónoma de México
Temixco, Morelos 62580, MEXICO. E-mail: pkn@cie.unam.mx

ABSTRACT

Ag$_2$S thin films of 90 nm to 300 nm in thickness were deposited at 70°C on glass substrates immersed in a bath mixture containing silver nitrate, sodium thiosulfate and dimethylthiourea. When the films are heated in nitrogen at temperatures 200°C to 400°C, crystallinity is improved and XRD pattern similar to that of acanthite is observed. These films possess electrical conductivity of 10^{-3} (ohm cm)$^{-1}$, are photoconductive and exhibit an optical band gap of 1.36 eV. When Ag$_2$S thin film is deposited over a thin film of Bi$_2$S$_3$, also obtained by chemical bath deposition from bismuth nitrate, triethanolamine and thioacetamide, and heated at 300°C to 400°C in nitrogen, a ternary compound, AgBiS$_2$ is formed. This material has an electrical conductivity of 5x10^{-5} (ohm cm)$^{-1}$, is photoconductive and possesses optical band gap 0.95 eV.

INTRODUCTION

Ag$_2$S in bulk form is reported to possess a direct optical band gap of 1.0 eV [1], and hence it is considered as a good solar radiation absorber.

In 1967 Kitaev *et al* [2] reported the conditions to obtain silver sulfide films over solid surfaces. In this work the authors mentioned the ability of thiourea to form insoluble metal chalcogenide salts. Other authors like Varkey [3], Lokhande [4], Dhumure and Lokhande [5, 6] and Grozdanov [7] reported on chemically deposited Ag$_2$S thin films using different complexing agents and sulfide ion sources. Depending on the deposition bath and conditions of deposition, distinct band gaps and electrical conductivities have been ascribed to the films by these authors.

In this work we report on Ag$_2$S thin films deposited from chemical baths containing silver thiosulfate complex and dimethylthiourea as a source of sulfide ions. This technique is very similar to that reported for the deposition of CuS thin films [8]. The variation in the crystalline nature of the film with annealing, optical and electrical properties are reported. The deposition of Ag$_2$S thin films on a chemically deposited Bi$_2$S$_3$ thin film and the production of a ternary compound upon heating is also reported. Since Ag$_2$S, Bi$_2$S$_3$ as well as ternary compounds derived from these are relatively non toxic, we consider them as relevant for solar cell technologies.

EXPERIMENTAL DETAILS

Bath composition

In the present study, corning microscope glass slides (75 mm x 25 mm x 1 mm) were coated with a thin film of ZnS of thickness < 20 nm to serve as a substrate film by the following procedure [9].

ZnS deposition bath: 0.62 ml of 1M $ZnSO_4$, 0.68 ml of 50% triethanolamine (TEA), 0.55 ml of buffer solution pH10, 0.25 ml of 1M thioacetamide (TA) and deionized water to complete 100 ml. Deposition was done at room temperature ($22^{\circ}C$ to $25^{\circ}C$) for 24 h.

Ag2S deposition bath: 25 ml of 0.1M $AgNO_3$, 9 ml of 1M $Na_2S_2O_3$, 10 ml de 0.5M dimethilthiourea (DMTU) and deionized water to complete 100 ml. Deposition was made on ZnS coated substrate at $70^{\circ}C$ during 3 h – 7 h. At 1 hour, the deposition of Ag_2S thin film was noticed; in 7 hours the deposition was completed. The glass slides were removed, rinsed in distilled water and dried in air. The films were yellow/brown in transmission and specularly reflecting.

Bi2S3 deposition bath [10]: 5 ml of 0.5M $Bi(NO_3)_3$, 4 ml of 50% TEA, 2 ml of 1M TA and deionized water to complete 100 ml. Deposition was done at $35^{\circ}C$ during 4h.

Bi2S3-Ag2S thin films: The Bi_2S_3 coated substrates were introduced into a freshly prepared Ag2S bath. For the film reported here, Bi_2S_3 film was deposited for 4 h at $35^{\circ}C$ (approximated thickness 190 nm) and Ag_2S thin film for 7 h at $70^{\circ}C$ (300 nm).

The thin films were then annealed in an oven at 100 mtorr nitrogen atmosphere at $300^{\circ}C$ for 1 hour.

Characterization:

The measurements of the thickness were done in a Alfa-step 100 unit (Tencor Instruments, CA). X-ray diffraction patters were recorded using Cu-K_{α} radiation on a Rigaku D-Max diffractometer. The optical spectra of the films were recorded on a Shimadzu UV-3101PC UV-VIS-NIR scanning spectrophotometer. For the electrical measurements (current vs. time), data were recorded on a computerized system using a Keithley 619 electrometer and a Keithley 230 programmable voltage source by applying a bias 1 V across coplanar silver print electrodes (5 mm x 5 mm). The samples were illuminated with 2 kW m^{-2} tungsten-halogen light. The measurements of the current were recorded for the initial 20 s in the dark, followed by 20 s under illumination and finally for 20 s in the dark.

RESULTS

The thin films of Ag_2S and Bi_2S_3 are amorphous as prepared. The XRD patterns of the samples annealed at $300^{\circ}C$, show the presence of Ag_2S (JCPDS 14-0072, monoclinic), figure 1; Bi_2S_3 (JCPDS 17-0320, orthorhombic), figure 2 and $AgBiS_2$ (JCPDS 21-1178, cubic) figure 3, respectively.

In figure 4 are shown the optical transmittance and reflectance, as well as the corrected transmittance, calculated following equation 1. The values are given in percentages.

$$T_{Corr} = \frac{T}{100 - R} x100 \tag{1}$$

The T_{Corr} spectra show absorption thresholds, from which we can deduce approximate values for the optical band gaps.

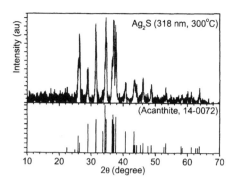

Figure 1. XRD pattern for a sample of Ag_2S annealed at 300°C.

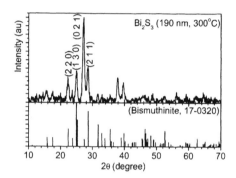

Figure 2. XRD pattern for a sample of Bi_2S_3 annealed at 300°C.

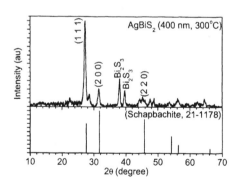

Figure 3. XRD pattern for a sample of $AgBiS_2$ formed by heating Bi_2S_3-Ag_2S at 300°C.

From T_{Corr}, the optical absorption coefficient (α, in cm^{-1}) of the material of the film is obtained using equation 2:

$$\frac{T_{Corr}}{100} = e^{-\alpha d} \tag{2}$$

where d is the thickness of the film in cm.

In general α depends on photon energy (hν) according to equation 3:

$$\alpha h\nu = B\left(h\nu - E_g\right)^n \tag{3}$$

where B is a constant and E_g is the optical band gap. The value of n depends on the type of transition (direct: n = 1/2, direct forbidden: n = 3/2, indirect: n = 2, indirect forbidden: n = 3) [11].

The straight lines observed for $(\alpha h\nu)^{2/3}$ versus hν plots, shown in figure 5 permit us to conclude that the samples have direct band gaps, with forbidden transitions: E_g = 1.36 eV for Ag$_2$S, 1.28 eV for Bi$_2$S$_3$ and 0.96 eV for AgBiS$_2$.

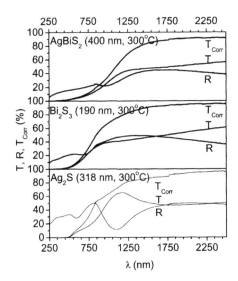

Figure 4. T, R and T_{Corr} plots of the Ag$_2$S, Bi$_2$S$_3$ and AgBiS$_2$ thin films.

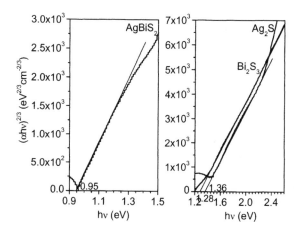

Figure 5. Optical absorption plots for the Ag_2S, Bi_2S_3 and $AgBiS_2$ thin films.

Figure 6 shows the photocurrent response curves of the thin films annealed at 300°C. In all the films some degree of photosensitivity is observed. From the data recorded, we can obtain the conductivity both in the dark (σ_D) and under illumination (σ_L). Table 1 shows these values along with the optical band gap.

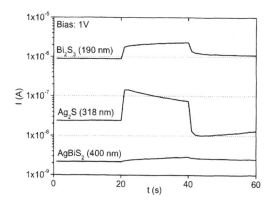

Figure 6. Photocurrent response curves of the Ag_2S, Bi_2S_3 and $AgBiS_2$ thin films.

Table 1. Conductivity of the thin films in dark and under illumination

	Ag$_2$S	Bi$_2$S$_3$	AgBiS$_2$
E$_g$ (eV)	1.36	1.28	0.95
σ_D (Ω cm)$^{-1}$	7.56x10^{-4}	4.75x10^{-2}	5.16x10^{-5}
σ_L (Ω cm)$^{-1}$	3.17x10^{-3}	1.15x10^{-1}	6.28x10^{-5}

CONCLUSIONS

In this paper we report Ag$_2$S thin films obtained by chemical bath deposition using thiosulphate as complexing agent and dimethilthiourea as sulfide-ion source. The thin films of Ag$_2$S, Bi$_2$S$_3$ and AgBiS$_2$ show good uniformity and adherence on ZnS coated glass substrates. Their optical energy band gap (0.96 – 1.36 eV) is in the desired interval to be used as absorber materials in solar cell structures. The electrical conductivity 10^{-5} to 10^{-2} (Ωcm)$^{-1}$ is in the range required for absorber materials to effectively absorb solar radiation in the depletion region of a junction. The as-prepared samples are amorphous, but when annealed at 300°C, in nitrogen atmosphere (300 mTorr) during 1 h, crystallinity is observed. The structures correspond to those reported for acanthite (Ag$_2$S), bismuthinite (Bi$_2$S$_3$) and schapbachite (AgBiS$_2$).

ACKNOWLEDGMENTS

We acknowledge the financial support received from CONACYT 34638-E. A. Nuñez acknowledges the financial support given by DGEP-UNAM.

REFERENCES

1. O. Madelung (editor) in *Semiconductors other than Group IV Elements and III-V Compounds* (Ed. Springer-Verlag Berlin Heidelberg, 1992), p. 13.
2. G.A. Kitaev, T.P. Bolshchicova and T.A. Ustianzeva (in russian), *Inorganic Materials* **3** (6), 1080-1082 (1967).
3. A.J. Varkey, *Solar Energy Materials* **21**, 291-296 (1991).
4. C.D. Lokhande, *Materials Chemistry and Physics* **27**, 1-43 (1991).
5. S.S. Dhumure and C.D. Lokhande, *Materials Chemistry and Physics* **27**, 321-324 (1991).
6. S.S. Dhumure and C.D. Lokhande, *Materials Chemistry and Physics* **28**, 141-144 (1991).
7. I. Grozdanov. *Semicond. Sci. Technol.* **9** (6-9), 1234-1241 (1994).
8. M.T.S. Nair, L. Guerrero and P.K. Nair. *Semicond. Sci. Technol.* **13**, 1164-1169 (1998).
9. O.L. Arenas, M.T.S. Nair and P.K. Nair. *Semicond. Sci. Technol.* **12**, 1323-1330 (1997).
10. M.T.S. Nair and P.K. Nair. *Semicond. Sci. Technol.* **5**, 1225-1230 (1990).
11. R.A. Smith in *Semiconductors*, 2nd ed. (Cambridge University Press, Great Britain, 1978) pp. 309-326.

Interaction between Poly(vinylidene fluoride) Binder and Graphite in the Anode of Lithium Ion Batteries: Rheological Properties and Surface Chemistry

Mikyong Yoo, [a] Curtis W. Frank, [a,b]
[a]Department of Materials Science and Engineering and [b]Department of Chemical Engineering, Stanford University, Stanford, CA 94305

ABSTRACT

We describe here the interaction of poly(vinylidene fluoride) (PVDF) with graphite based on the rheological behavior of the slurries and the surface morphology of PVDF in the final composite anode. The rheological properties of slurries show the interaction between graphite and PVDF with the viscosity varying over six orders of magnitude for different graphite particles. We correlated the suspension viscosity with the final film properties. The homogeneity of the PVDF distribution in the final composite film increases as the slurry viscosity increases. The interaction between graphite and PVDF is altered through the chemical properties of polymer such as molecular weight and functionality, leading to an improved morphology of PVDF.

INTRODUCTION

In spite of the recent advances in lithium ion rechargeable batteries, there remain challenging problems related to the fabrication of the anode and cathode, which are both composite materials. The anodes consist predominantly of graphite particles (90-98% by weight) bound together by a polymeric binder such as poly(vinylidene fluoride) (PVDF). Although PVDF is the most promising, other fluorinated binders have been used, including poly(tetrafluoro ethylene) (PTFE) [1,2] as well as non-fluorinated polymers such as poly(ethylene propylene diene) (EPDM) [3,4]. The polymeric binder, which is necessary to provide sufficient mechanical strength to the electrodes, can influence the electrochemical reaction between lithium ions and electrolyte that takes place at the carbon surface, since it is known that the details of the surface chemistry and the morphology of the carbon play important roles in SEI formation [3,5], and that the binder influences heat generation in the lithium ion batteries [4,6]. Moreover, it has been reported that up to 70% of the graphite surface was covered by PVDF binder even though the concentration of binder was kept relatively low [7]. Therefore, it is important to understand and control the binder/particle interactions and the relationship between the processing parameters for the complex fluid suspension and the final composites.

In this paper, we will examine binder interaction with graphite using different types of graphite mixed with PVDF binder. The PVDF binder/graphite interaction is manifested in the rheological properties in the high-solids-content slurries. Because such slurries are complex fluids with their rheological properties important for determining the stability of the suspension, their study is closely related to the processing of composite electrodes. And we will present the morphology of binders on the final composite anode and correlate the interaction between graphite and binder of the slurries with the surface properties of final composite films. This interaction and the surface distribution of PVDF are altered through chemical properties of PVDF such as molecular weight and modified functionality.

EXPERIMENTAL DETAILS

We used eight types of carbon materials, and three different PVDF, as shown in Table I, to investigate the effect of different surface area and nature of carbon and the chemical properties of PVDF on the interaction between carbon particles and PVDF. Anodes were prepared by mixing carbon slurries that contained the carbon particle, 1-methyl-2-pyrrolidinone (NMP, 99.9+%, Mitsubishi Chemical Co.) as a carrier, and 10 wt% solution of PVDF binder in NMP solvent. The solid concentration of the slurries was 40 wt%. We spread the slurry using the doctor-blade method on a sheet of copper foil and dried it in an oven at 83 °C in air for 1 hour to form 95/5 wt% of graphite/PVDF composite anode.

Dynamic viscosities of the slurries were measured at 23°C by a stress-controlled rheometer (DSR, Rheometric Scientific) for MPG-V2 and MBC-N and a strain-controlled rheometer (Dynamic analyzer RDA II, Rheometric Scientific) for the other samples. Two rheometers were required because the viscosities of MPG-V2 and MBC-N slurries were too low for the range of the strain-controlled rheometer. We used a 50 mm parallel plate with a 1 mm gap, and samples were prepared just before the experiments. The viscosity was measured as a function of the frequency in the range of 0.1 and 100 rad· sec^{-1} at the maximum strain amplitude of the linear viscoelastic region.

The spatial distribution of PVDF on the graphite particles was evaluated by mapping the fluorine of PVDF using electron probe X-ray microanalysis (EPMA) (JXA-733, JEOL) with 50 nA of current. In the case of different PVDF samples, energy dispersive spectroscopy (EDS, JSM-5600LV, JEOL) was used with a gun voltage of 10 kV. We determined the degree of homogeneity of PVDF by quantifying the image of fluorine dots using a spatial autocorrelation function. From the characteristics of the autocorrelation function, we were able to calculate the area of a fluorine dot cluster and the spacing between clusters. From these values, we derived the degree of homogeneity.

Table I. Information on carbon particles and PVDF used for the experiments.

	Manufacturer	Type of carbon	Average particle size(μm)	BET surface area (m^2/g)
MPG-V2	Mitsubishi Chemical Co.	Synthetic graphite	18.4	2.8
MBC-N	Mitsubishi Chemical Co.	Amorphous carbon	18.0	5.0
MCMB	Oosaka Gas Chem.	Synthetic graphite	7.8	3.3
SFG44	Timcal Co. Ltd.	Synthetic graphite	22.0	5.0
SFG75	Timcal Co. Ltd.	Synthetic graphite	27.0	3.5
KS15	Timcal Co. Ltd.	Synthetic graphite	7.7	12.0
SFG15	Timcal Co. Ltd.	Synthetic graphite	8.1	8.8
KS6	Timcal Co. Ltd.	Synthetic graphite	3.3	20.0
	Manufacturer	Functionality	Molecular weight	H-H defects(%) (^{19}F NMR)
KF1300	Kureha	N/A	~ 350,000	4.22 ($\pm$ 0.53)
MKB212A	Atofina	-OH (a little – COOH)	~ 350,000	5.54 ($\pm$ 0.67)
Kynar301F	Atofina	N/A	~ 500,000	6.55 ($\pm$ 1.24)

RESULTS AND DISCUSSION

Rheological properties of slurries

To investigate the interaction between carbon particles and PVDF before doctor-blading, we measured the viscosity of the slurries. Figure 1 shows the frequency dependence of the real viscosity, which shows over six orders of magnitude difference. We used the real viscosity here (loss modulus divided by frequency) instead of the dynamic viscosity because the storage moduli of MPG-V2 and MBC-N showed negative values in the high frequency region, which does not have any physical meaning. For MPG-V2 and MBC-N, the real viscosities are nearly independent of frequency on a log-log scale, which indicates that these behave as Newtonian interfaces. On the other hand, the viscosities of the others decrease linearly as frequency increases. This shear thinning is typical non-Newtonian behavior. In concentrated particle/polymer suspensions, the shear thinning occurs due to structural breakdown. When a polymer adsorbs at a solid-liquid interface, a portion of the polymer loop or tail may be extended from one particle to another particle. Therefore, flocculation of suspensions may be modeled by a bridging process [8]. In a developed three-dimensional network of flocs, structural breakdown is induced with increasing shear rate, leading to shear thinning behavior.

High BET surface area of graphite particles generally causes the high interaction between graphite and PVDF through more adsorption area. The influence of reactive sites on the interaction can be examined by using a modified PVDF. When we used a PVDF modified by the hydroxyl functional group, which has the same molecular weight as that of KF1300, the dynamic viscosity increases in terms of the same graphite as shown in Fig. 2. This indicates that the altered interaction between graphite and PVDF through the hydroxyl functional group enhances the adsorption of PVDF onto the graphite particles. The higher molecular weight PVDF, Kynar301F, causes the viscosities to be increased because high molecular weight polymer can bridge more particles together with longer chains, leading to a more flocculated particle structure. PVDF is adsorbed onto the surfaces of graphite through hydrogen bonding [9] and physisorption, and loops and tails can bridge more than one particle together, leading to aggregated particles with high viscosity.

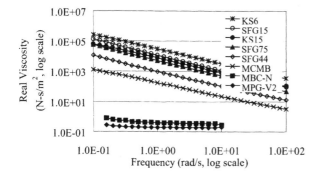

Figure 1. Dependence of real viscosity on frequency for slurries at room temperature. Viscosities are significantly different over six orders of magnitude. Whereas MPG-V2 and MBC-N show characteristics of a Newtonian fluid, other samples show non-Newtonian shear thinning.

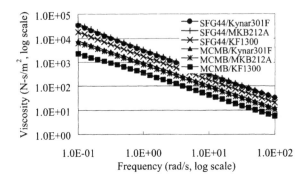

Figure 2. Viscosity influenced by the chemical properties of PVDF in terms of graphite particles. The high molecular weight PVDF (Kynar301F) and the modified one (MKB212A) show a higher viscosity.

Surface distribution of PVDF

The location and morphology of PVDF on the carbon can be observed by detecting fluorine with EPMA because PVDF has a repeating unit of $-(CH_2-CF_2)-$. Since the depth of field of EPMA is about 1 μm, we cannot detect the *exact* surface morphology of PVDF, but rather we determine a *near*-surface morphology. Figure 3 (a)-(f) show fluorine dot mappings of the composite films with different carbon materials. These images are shown in the order of increasing slurry viscosity (see Fig. 1), illustrating that the distribution of PVDF in the final composite film becomes more homogeneous with increasing viscosity. We used different magnifications to minimize the influence of carbon particle size on the images of apparent polymer distribution. We also used the normalized value considering the magnification and the average particle size. We quantified the degree of homogeneity by using the autocorrelation function, C(R), which describes the spatial correlation of the density of dots in an image and is

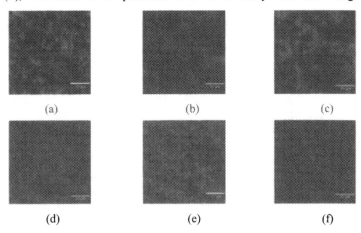

Figure 3. Fluorine dot mapping of (a) MBC-N, (b) MCMB, (c) SFG44, (d) KS15, (e) SFG15, and (f) KS6. These images are shown in the order of increasing viscosity. 120 pixel corresponds to 20 μm for (a) and 10 μm for (b), (c), (d), and (f). For (e), 116 pixel corresponds to 10 μm.

defined as the average normalized by C(0) [10],

$$C(R) = \frac{<[\rho(r) - \rho_o][\rho(r-R) - \rho_o]>}{C(0)} \qquad (1)$$

where ρ is a dot density distribution, and the angular brackets indicate averaging over all coordinates r. When r is equal to 0, there is a strong self-correlation peak corresponding to each image correlating with itself. When r is equal to the mean displacement, there is a second peak called a positive displacement peak, which corresponds to the first image correlating with the second image. From this characteristic of the autocorrelation function, we can evaluate the area of a cluster and the spacing between clusters. The two-dimensional image was transformed to the one-dimensional spatial peak as shown in Fig. 4(a). The area under the first peak indicates the size of a cluster and the distance to the second peak corresponds to the radial average spacing between two clusters. We plotted the spacing versus the average size of clusters of each sample in Fig. 4(b); small cluster size and reduced distance between clusters of dots indicate homogeneous distribution. The data approaches a small cluster size and spacing as the viscosity increases. From these data, we relate the processing condition such as viscosity to the surface properties of final films.

　　Figure 5 shows the effect of the molecular weight and functionality of PVDF on the surface distribution of PVDF. We plotted the values for each graphite and PVDF and connected points in terms of PVDF. MKB212A shows an improved homogeneity compared to KF1300 in all carbon particles through enhanced interaction with graphite particles. In the case of Kynar301F, better homogeneity of PVDF is detected for some carbon samples, which can be caused by long chains of high molecular weight PVDF that can be more adsorbed onto graphite particles. The homogeneity range of Kynar301F PVDF is small compared to other PVDF materials because of the aggregation of long chains of PVDF after the adsorption of PVDF onto the surface of

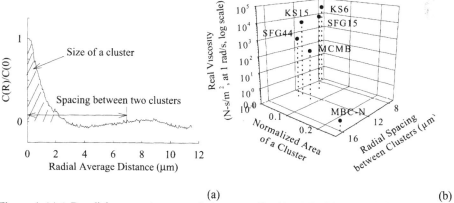

(a)   (b)

Figure 4. (a) 1-D radial averaged autocorrelation normalized by C(0). The area of the first peak indicates the size of a cluster and the distance to the second peak corresponds to the radial average spacing between two clusters. (b) 3-D graph showing the relationship between the degree of homogeneity and real viscosity. As the viscosity increases, the homogeneity becomes improved.

171

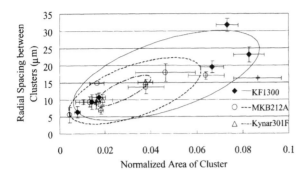

Figure 5. Radial spacing between clusters of fluorine dots versus normalized area of cluster in terms of PVDF. Kynar301F and MKB212A show an improved homogeneity compared to KF1300.

graphite. The Kynar301F and MKB212A show higher interaction with graphite particles, determined by the rheological properties, leading to a more homogeneous distribution of PVDF on the surface of graphite particles.

CONCLUSIONS

In this paper, we have demonstrated that the interaction between graphite and PVDF in slurries depends on the nature of graphite particles and the chemical properties of PVDF and can be correlated with the surface properties of the final composite film. This study of particle/PVDF interaction is important in understanding and controlling the final film properties, which could contribute to the improvement of the electrochemical efficiency in lithium ion batteries.

ACKNOWLEDGMENTS

This work was supported by Mitsubishi Chemical Corporation.

REFERENCES

1. G. B. Li, R. J. Xue, and L. Q. Chen, *Solid State Ionics* **90**, 221 (1996).
2. W. F. Liu, X. J. Huang, G. B. Li, Z. X. Wang, H. Huang, Z. H. Lu, R. J. Xue, and L. Q. Chen, *J. Power Sources* **68**, 344 (1997).
3. O. Chusid, Y. E. Ely, D. Aurbach, M. Babai, and Y. Carmeli, *J. Power Sources* **43**, 47 (1993).
4. M. N. Richard and J. R. Dahn, *J. Power Sources* **83**, 71 (1999).
5. E. Peled, C. Menachem, D. BarTow, and A. Melman, *J. Electrochem. Soc.* **143**, L4 (1996).
6. H. Maleki, G. P. Deng, I. KerzhnerHaller, A. Anani, and J. N. Howard, *J. Electrochem. Soc.* **147**, 4470 (2000).
7. K. A. Hirasawa, K. Nishioka, T. Sato, S. Yamaguchi, and S. Mori, *J. Power Sources* **69**, 97 (1997).
8. T. W. Healy, *J. Colloid Sci.* **16**, 609 (1961).
9. A. Biswas, R. Gupta, N. Kumar, D. K. Avasthi, J. P. Singh, S. Lotha, D. Fink, S. N. Paul, and S. K. Bose, *App. Phys. Lett.* **78**, 4136 (2001).
10. G. R. Strobl and M. Schneider, *J. Polym. Sci. B* **18**, 1343 (1980).

Mat. Res. Soc. Symp. Proc. Vol. 730 © 2002 Materials Research Society

PZT AND ELECTRODE ENHANCEMENTS OF MEMS BASED MICRO HEAT ENGINE FOR POWER GENERATION

A.L. Olson, L.M. Eakins, B.W. Olson, D.F. Bahr, C.D. Richards, R.F.Richards
Mechanical and Materials Engineering, Washington State University, Pullman, WA

ABSTRACT

The P3 Micro Heat Engine relies on a thin film PZT based transducer to convert mechanical energy into usable electrical power. In an effort to increase process yield for these devices, PZT adhesion, cracking, and thickness effects were studied. Also, to increase power generation, novel PZT etching patterns were used to improve the strain at failure. AFM and AES were used on sputtered Ti/Pt bottom electrodes to compare roughness, grain size, and diffusion for annealing temperatures between 550 and 700 °C. For an optimized bottom electrode, process yield for various sized top electrodes were then studied for PZT thickness between 0.54 and 1.62 μm. Analytic strain calculations in pressurized square membranes, showing high strain at half the lateral distance from each corner of the membrane, were then used as a qualitative model for reducing stress concentrations. Two PZT etching geometries on 2.3 μm thick Si/SiO2 membranes, with 1.5-3.5 mm side-lengths, were examined and one was used to increase the strain at failure by at least 40%. Integrating improvements in process yield and strain at failure, single PZT based MEMS devices capable of generating power of up to 1 mW and in excess of 2 volts have been demonstrated operating at frequencies between 300 and 1,100 Hz.

INTRODUCTION

The P3 Micro Heat Engine is a small mm-scale power generation device relying on two-phase fluid expansion to deflect a μm-scale thin-film piezoelectric membrane [1]. In this case, lead zirconate titanate (PZT) is used. Two main sections comprise the heat engine, a heat source and a generator (the piezoelectric transducer). Figure 1 shows the cross section of such a device. Heat is supplied to the engine through the source causing pressure to develop in the engine cavity from evaporation. This pressure deflects the generator and the resulting strain in the piezoelectric film causes a voltage to develop across the top and bottom electrode, effectively a d_{31} mode of the piezoelectric effect. Voltage cycles are then used to drive a charging circuit that generates power.

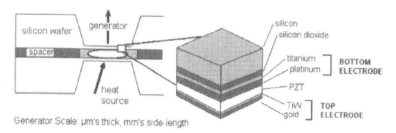

Figure 1. The P3 Micro Heat Engine and generator (piezoelectric transducer)

PZT adhesion, cracking, and electrical defects due to the bottom electrode structure degrade the performance of piezoelectric devices and process yield. For this reason the bottom electrode must have good adhesion to the silica substrate as well as good surface morphology and chemistry to support PZT with large residual tensile stress. Pt is commonly used as the bottom electrode for its temperature stability as well as its promotion of high quality perovskite PZT crystal growth [2]. Pt has also been chosen over lower resistance metals for its lack of interaction with dielectric materials in subsequent high temperature processing steps. A 5-15 nm layer of Ti is used to promote adhesion.

One of the problems with using Ti as an adhesion layer is its diffusion through Pt upon annealing or subsequent high temperature processing. Diffusion has been shown to reduce the sheet resistance of the bottom electrode due to formation of Pt-Ti alloys and TiO_x [3]. Also, diffusion of Ti may create Ti encapsulated hillocks [4]. Intermediate adhesion layer diffusion has also been linked to changes in PZT structure [5].

The formation of hillocks in the Pt layer can be minimized with appropriate titanium to platinum ratios for a given annealing time and temperature. Larger ratios of Pt to Ti have been shown to reduce the number of hillocks but increase their size, while small ratios increase the number and size of hillocks [3]. Intermediate ratios provide the best results. The formation of stress relieving hillocks may also be restricted by spin coated PZT using a sol-gel method.

Through thickness fracture in sol-gel derived thin film PZT membranes also impacts the performance of piezoelectric pressure transducers. This failure mechanism is very important to the useful operation of these devices for the maximum strain, voltage, and deflection can be physically limited by fracture. The strain in pressurized square membranes where deflections are large with respect to film thickness can be calculated to account for large scale out of plane deflections with the addition of non-linear terms by [6]

$$\varepsilon_x = \frac{du}{dx} + \frac{1}{2}\left(\frac{dw}{dx}\right)^2 ; \; \varepsilon_y = \frac{dv}{dy} + \frac{1}{2}\left(\frac{dw}{dy}\right)^2 ; \; \gamma_{xy} = \frac{du}{dy} + \frac{dv}{dx} + \frac{d^2w}{dxdy} \qquad (1)$$

where ε is the strain, u and v are in plane displacement, and w is out of plane displacement. An energy minimization routine can then used to find functional forms of the displacement field [7]. This entails balancing the strain energy in the membrane with the potential energy from the applied pressure and minimizing the total potential energy with respect to unknown "shaping" parameters in the displacement field.

The generalized pressure-deflection relationship typically used in the energy minimization routine is composed of solutions for two limiting cases [8]. The linear term was obtained for small deflections in a residual stress dominated regime, and the second term was obtained for large deflections where residual stress can be neglected. For the large deflections we attain, strain calculations were based on negligible residual stresses contributions [8]. Pressure-deflection measurements (p and h respectively) can also be used to calculate modulus (E) and residual stress (σ_0) by

$$p = c_1 \sigma_0 t \frac{h}{a^2} + c_2 \frac{Et}{(1-v)} \frac{h^3}{a^4} \qquad (2)$$

for a side-length of 2a, film thickness t, and Poison ratio v when given c_1 and c_2 (3.393 and 1.981-0.585v respectively). Strain calculations in square membranes show maximum values at a lateral distance a from each corner of the membrane, as seen in Figure 2a.

EXPERIMENTAL PROCEDURES

Device fabrication is described elsewhere [9]. To summarize, a 150 nm low temperature oxide was grown on 2.3 μm thick square Si membranes, with 1.5, 2.5, and 3.5 mm side-lengths made by preferentially etching (001) silicon. Ti/Pt bottom electrodes were then DC magnetron sputtered on wafers without membranes to study yield and thickness effects. A 12.5 nm thick Ti film was sputtered at 1.7 mTorr. Without breaking vacuum, Pt was sputtered at 14 mTorr to deposit a 175 nm thick film. Bottom electrodes were then annealed at 550, 600, 650, and 700 °C for 10 minutes. 52/48 PZT with 10% excess Pb content was spin coated at 3,000 rpm for 30 seconds to produce 90 nm thick layers that were individually pyrolyzed at 375 °C for 3 minutes. After three or four layers, the PZT was crystallized at 700 °C for 10 minutes. Successive PZT layers were applied until the desired total thickness was achieved. A top electrode of TiW/Au was then sputtered at 3.8 mTorr, patterned using photolithography, and etched. TiW/Au is chosen for its oxidation resistance during poling and ease of processing.

Roughness and grain size were then measured on bottom electrodes annealed at 550, 600, 650, and 700 °C using a Park Autoprobe CP AFM in contact mode. Eight scans of 2, 3 and 5 μm were used to obtain average values for each film. Grain size calculations were made using the Heyn intercept method. Vickers indents into samples with PZT were then used as a qualitative adhesion test by spalling area. A Phi 680 AES was used to determine chemistry at the platinum interface before and after PZT deposition and species profiles begun at the Pt surface where PZT was spalled during indentation. Yield was found by measuring the resistance of transducer stacks with square electrodes of side-length: 0.5, 1.5, 2.0, 2.5, and 3.0 mm. The transducer was considered successful if the resistance exceeded 20 MΩ.

Based on analytic strain calculations, two etching geometries were selected to study the effect of removing PZT from 3 or 4 high strain locations (Figure 2c and 2b respectively). The transducers contained 1.08 μm of PZT between the top and bottom electrodes. On the same wafer, etching geometries were patterned and PZT was wet etched from exposed areas. In total membranes were un-etched, 3-side etched, and 4-side etched with approximately 15, 25, and 50% of the PZT removed from each membrane. It should be emphasized that the top electrode area covers the same proportion of the membrane for 1.5, 2.5, and 3.5 mm side-lengths, and the proportion of PZT removed has 3 values for each etching geometry.

Membranes with the various etching geometries were then poled at room temperature for 30 seconds with 15 V and tested using a dynamic bulge system with laser interferometry[10]. To summarize, an actuator is used to force the diaphragm, water, and membrane system. At

Figure 2. (a) Typical strain profile in a square membrane, (b) PZT etching geometry to relieve 4 high strain regions, (c) PZT etching geometry to relieve 3 high strain regions

frequencies between 300 and 1,100 Hz, the system would resonate producing deflections and voltages from the transducer. An important note is that the time response of the PZT dictates the use of a dynamic system, and testing at the resonant frequency is only for obtaining large deflections from this system. Voltage and deflection were simultaneously measured, but dynamically obtained deflections do not correspond to equation (2).

RESULTS AND DISCUSSION

Table 1 illustrates the trends observed in roughness and grain size for bottom electrodes annealed at various temperatures. The bottom electrode RMS roughness was found to spike at 78 Å at or below 550 °C and decrease to a value similar to the as deposited, 54 Å, with increased temperature. At the same time, the PZT deposited on these samples attains a smooth, crack free surface at intermediate roughness values. The grain size was found to increase with temperature, and AES depth profiles showed no titanium accumulation at the interface and little or no diffusion before or after PZT deposition. Peak to valley roughness, a measure of hillock formation, followed a similar trend as RMS roughness.

Figure 3a shows qualitatively that PZT adhesion to Pt appears to improve significantly with increased annealing temperature. Vickers indents at 500 g were made into PZT films on bottom electrodes and the area of PZT that spalled off was measured. The average and range of PZT-spall area decreases with annealing temperature. The trend in PZT adhesion may correlate with grain size, but similar roughness, between the as deposited and 700 °C annealed bottom electrodes, produces significantly different adhesion. PZT adhesion was not affected by titanium diffusion since both good and poor adhesion occurred without Ti diffusion.

Figure 3b demonstrates the trend in device yield for a bottom electrode suitably annealed at 650 °C for 10 minutes to attain crack free PZT with good adhesion. Each data point represents the average yield from 40 to 50 devices with the same electrode area. Larger electrodes were found to produce a disproportionate increase in yield with thicker PZT. Although some hillock formation is observed when bottom electrodes are annealed at 650 °C, hillock and intrinsic defects appear to significantly diminish with increased PZT thickness.

The voltage output from membranes with 1.5, 2.5, and 3.5 mm side-lengths is plotted in Figure 4a as a function of strain as calculated using the energy minimization method assuming negligible residual stress [8]. The figure also represents data obtained when PZT was removed from 3 high strain locations, 4 high strain locations, and when PZT was not removed. The linear relation between voltage and strain, despite different etching geometry and size, suggests that neglecting the residual stress term in (2) is a reasonable assumption for the strain calculations. Static pressure deflection curves also show only a slight shift (5% at 60 microns) in compliance.

| Temp.(°C) | Roughness (Å) | | Grain Size (nm) | Observations |
	RMS	Peak to Valley		
23	54	505	62	PZT delamination
550	78	808	64	Channel cracking
600	64	594	81	
650	63	573	111	
700	54	560	120	Channel cracking

Table 1. Roughness and grain size; 175:12.5 nm Pt:Ti bottom electrode annealed for 10 minutes

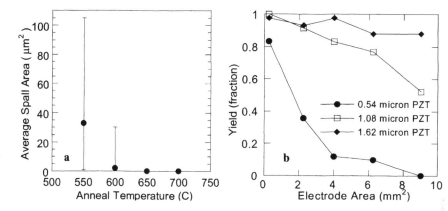

Figure 3. (a) Average spall area and total range from 10, 500 g, Vickers indents of PZT on annealed Pt electrodes; (b) Yield from 40-50 devices per point as a function of electrode area and PZT thickness on Pt annealed for 10 min. at 650°C

PZT etching was found to significantly improve the strain at failure for the piezoelectric transducers. Figure 4b represents the maximum strains attainable before membranes either broke or the dynamic response of the piezo-actuator (of the test apparatus) reached its forcing limitations. At large deflections, interference fringes were not resolvable and deflections were extrapolated from dynamic pressure-deflection curves. The maximum deflections and strains attainable increased with increasing size. Membranes with 1.5 mm side-length could only be forced to 0.07% strain at the center of the membrane before reaching the piezo-actuator forcing limits, and none of the membranes tested fractured regardless of being etched or not. Membranes with 2.5 mm and 3.5 mm side-lengths, however, failed at an average strain of 0.11% and 1.47 V, for 5 out of 6 membranes tested with PZT both etched from 3 high strain locations or not etch (the blanket PZT film). This corresponds to a "global" maximum strain, located at the center of the side-length of the membrane, of 0.17%. When PZT was removed from 4 high strain locations, the membranes could be strained to the forcing limitations of the system without fracture. This corresponds to a strain at the center of the membrane of 0.15% and 2 V, which when cycled at 400 Hz correlates to 1 mW of power generation. Removing PZT from 4 high strain locations increased the voltage output by at least 40% without failure. A 0.17% strain at the center of the membrane should achieve 2.3 V. The voltage, strain, and failure did not change for membranes with various amounts of PZT removed for the etching geometries tested.

CONCLUSIONS

Bottom electrodes annealed at 650 °C for 10 minutes, with a Pt/Ti thickness ratio of 14, were found to produce crack free PZT and provide good interfacial adhesion. The adhesion did not correlate with titanium diffusion. Also, electrical defects in piezoelectric transducers were significantly reduced with increased PZT thickness thus improving device yield. Etching PZT from high strain locations on square membranes then improved the strain at failure by at least 40%. Batch fabricated PZT based MEMS devices capable of generating in excess of 2 V (and approximately 1 mW) have been demonstrated operating at 400 Hz.

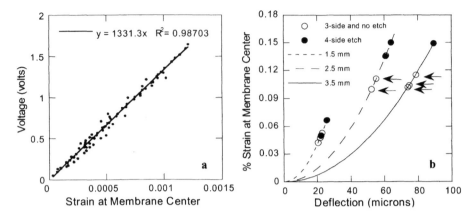

Figure 4. (a) Voltage/strain ratio of 1.5, 2.5, and 3.5 mm membranes with etched and un-etched PZT; (b) Strain improvements for membranes with PZT removed from 4 high strain locations, fractured membranes noted with arrows

ACKNOWLEDGMENTS

This research was supported by NSF under grant #9980837. The AES characterization described in this paper was performed at the Environmental Molecular Sciences Laboratory, a national scientific user facility sponsored by the Department of Energy's Office of Biological and Environmental Research and located at Pacific Northwest National Laboratory. The authors would like to thank Mr. J Skinner for operating the dynamic bulge system.

REFERENCES

1. C. Xu, J. Hall, C. Richards, D. Bahr, and R. Richards, ASME IMECE MEMS Symposium MEMS **2**, 261 (2000).
2. San-Yuan Chen, I-Wei Chen, *J. Am. Ceram. Soc.*, **77**, 2332 (1994).
3. H.N. Al-Shareef, D. Dimos, B.A. Tuttle, M.V. Raymond, *J. Mater. Res.* **12**, 347 (1997).
4. G. Vélu, D. Rèmiens, *Journal of the European Ceramic Society* **19**, 2005 (1999).
5. Radosveta D. Klissurska, Thomas Maeder, Keith G. Brooks, Nava Setter, *Microelectronic Engineering* **29**, 297 (1995).
6. S. Timoshenko, S. Woinowski-Krieger, Theory of Plates and Shells, McGraw-Hill, New York (1982).
7. J.J. Vlassak, W.D. Nix, *J. Mater. Res.* **7**, 3242 (1992).
8. Robert J. Hohlfelder, Ph.D. Thesis, Stanford Unversity (1999).
9. D.F. Bahr, K.R. Bruce, B.W. Olson, L.M. Eakins, C.D. Richards, and R.F. Richards, Proceedings of the Materials Research Society **687**, 4.3.1 (2002).
10. J.D. Hall, N.E. Apperson, B.T. Crozier, C. Xu, R.F. Richards, D.F. Bahr, and C.D. Richards, Review of Scientific Instrumentation **73**, 2067 (2002).

Mat. Res. Soc. Symp. Proc. Vol. 730 © 2002 Materials Research Society   V5.21

Preparation of Doped Lanthanum Gallate Electrolyte for SOFC
by Pulsed Laser Deposition

Seiji Kanazawa[1], Takeshi Ito[1], Kenji Yamada[1], Toshikazu Ohkubo[1], Yukiharu Nomoto[1], Tatsumi Ishihara[2] and Yusaku Takita[2]

[1] Department of Electrical and Electronic Engineering, Oita University, 700 Dannoharu, Oita 870-1192, Japan

[2] Department of Applied Chemistry, Oita University, Oita University, 700 Dannoharu, Oita 870-1192, Japan

ABSTRACT

In this study, doped lanthanum gallate (LSGM with the composition $La_{0.9}Sr_{0.1}Ga_{0.8}Mg_{0.2}O_{3-}$, LSGMC with the composition $La_{0.8}Sr_{0.2}Ga_{0.8}Mg_{0.15}Co_{0.05}O_{3-}$) films for an electrolyte of the solid oxide fuel cell (SOFC) were prepared by pulsed laser deposition (PLD) technique. In the vacuum chamber, LSGM or LSGMC targets were set on the rotating target holder. A KrF excimer laser was introduced into the chamber at an incident angle of about 45 degree. The doped $LaGaO_3$ film was deposited onto NiO substrates without heating in argon ambient gas. The NiO substrate can be used directly as an electrode in the fabrication of the SOFC. The deposited LSGM films were characterized by X-ray diffraction (XRD), secondary ion mass spectroscopy (SIMS) and scanning electron microscopy (SEM). As-deposited films were amorphous. After post annealing at 1273K for 6-10 hours, crystalline $LaGaO_3$ was obtained. Films with thickness greater than several 10 m showed an uniform and dense morphology. No gas leakage was found using thick films, which is an important characteristic for an electrolyte in fuel cells. The composition of the deposited films was slightly different to that of the target.

INTRODUCTION

Sr-Mg-Doped $LaGaO_3$ oxide, with the composition $La_{1-x}Sr_xGa_{1-y}Mg_yO_{3-(x+y)/2}$ (LSGM) has shown good ion conductivity [1]. It has an ion conductivity of $0.1 Scm^{-1}$ at 1073K, comparable to that of yttria-stabilized zirconia (YSZ) at 1273K and it is a promising candidate for an electrolyte in solid oxide fuel cell (SOFC). Recently, the performance of the SOFC using $La_{0.8}Sr_{0.2}Ga_{0.8}Mg_{0.2-x}Co_xO_{3-}$ (LSGMC) as the electrolyte at 923K has been studied [2]. If high quality doped $LaGaO_3$ thin films can be prepared, an operation of SOFC at reduced temperatures may be possible. Conventional thin film preparation techniques such as sputtering, metal-organic chemical vapor deposition, and ion-beam deposition have been used to prepare a wide variety of thin films. However, the preparation of the thin films having the same stoichiometry as the target is very difficult. Pulsed laser deposition (PLD) is considered to be a promising

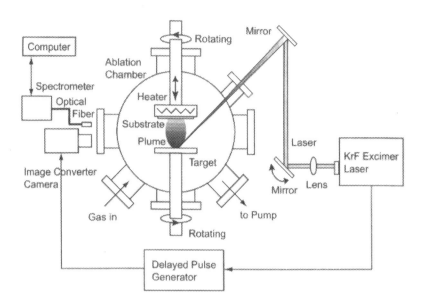

Figure 1. Pulsed laser deposition system.

method for the preparation of thin films of multicomponent stoichiometric composites. Up to now, various kinds of multicomponent films have been successfully prepared [3]. Recently, the deposition of doped $LaGaO_3$ thin films by using PLD technique has been attempted [4].

In this study, doped $LaGaO_3$ films, with the composition $La_{0.9}Sr_{0.1}Ga_{0.8}Mg_{0.2}O_{3-\delta}$ and $La_{0.8}Sr_{0.2}Ga_{0.8}Mg_{0.15}Co_{0.05}O_{3-\delta}$ were prepared by the PLD technique. Time resolved plume characteristics were measured, and film properties such a surface morphology, crystalline structure, and atomic component ratio were evaluated.

EXPERIMENTAL APPARATUS

Figure 1 shows a schematic diagram of the pulsed laser deposition and optical measurement systems used in this study. In the vacuum chamber (ϕ =400mm, stainless steel), the doped $LaGaO_3$ target (LSGM, 46mm in diameter, or LSGMC, 50mm in diameter) was set on the rotating target holder. A KrF excimer laser (Lambda Physik, COMPex 110, wavelength λ =248nm, pulse duration=30ns, maximum energy=350mJ) was introduced into the chamber at an incident angle of about $45°$ through a lens, mirrors and a quartz window. The laser beam with repetition rate of 10Hz was scanned on the surface of the target by using the rastering mirror. Accounting the laser beam size on the target, the laser fluence was estimated to be about $3J/cm^2$.

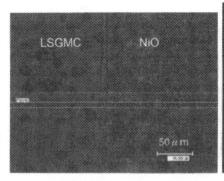 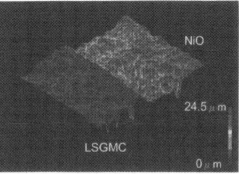

(a) Surface image   (b) 3D profile (Observed area is 300 m × 220 m)

Figure 2. Surface morphology of the LSGMC film deposited onto a NiO substrate (0.1Pa).

To avoid pitting the target during film deposition, the target was rotated at 7rpm.. The substrate is located 65-85mm above the target. The film was mainly deposited onto NiO substrates, in argon or oxygen in the pressure range from vacuum to 10Pa, at ambient temperature. The NiO substrates can be used directly as electrodes in the fabrication of the SOFC. To investigate the properties of ablation plasma, an image converter camera (Hadland photonics, 790) was used to obtain information about ablation plasma plumes. The excimer laser and the image converter camera were synchronously controlled by a delayed pulse generator. Light emitted from the plasma plume was also collected by a lens and imaged onto an entrance slit of a spectrometer (Ocean Optics, USB2000) through an optical fiber. The deposited LSGM and LSGMC films were characterized with measurements of X-ray diffraction (XRD), secondary ion mass spectroscopy (SIMS) and scanning electron microscopy (SEM). The surface morphology of the films was further evaluated by laser microscopy.

RESULTS AND DISCUSSION

Figure 2 shows the surface morphology of the LSGMC film investigated by laser microscopy. The films was deposited at 0.1Pa argon and annealed at 1373K for 6 h. The film thickness is 6.4 m and the deposition rate is about 60pm/pulse. As can be seen in figure 2, the film is not smooth due to the surface roughness of the NiO substrate. Especially, the holes can be seen both in the film and the substrate. In order to prepare the film without voids, the films were deposited at several pressures. As a result, a background pressure was found to have a significant influence on the surface morphology. The plasma plume was also affected by the background pressure. The velocity of the plume, which is deduced by time-resolved images of the plume captured by the image converter camera, varies from 1.5×10^4m/s in vacuum to 0.5×10^4m/s at 100Pa argon. It is considered that the dynamics of the energetic plume affects

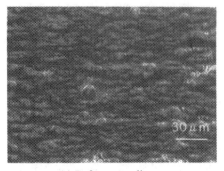

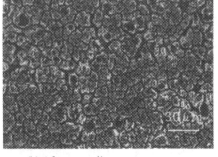

(a) Before annealing   (b) After annealing

Figure 3. SEM images of the LSGMC film deposited on NiO substrate (10Pa).

the properties of the films. The major emissive species in the plasma plume identified by optical emission spectra in the range of 300-800nm are La, La$^+$, Sr, Sr$^+$, Ga, Mg, LaO, MgO. Figures 3(a) and 3(b) show the SEM images of the surface of the film deposited at 10Pa before and after annealing, respectively. As can be seen, the trace of the pores in NiO substrate disappeared when the deposition was carried out at higher pressures. However, micro-cracks developed during annealing. In this case, the film thickness is 3.7 μ m and the deposition rate is about 34pm/pulse. In general, the deposition rate decreased with

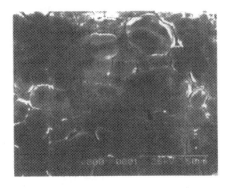

Figure 4. SEM images of the thick LSGM film deposited on NiO substrate (3Pa).

increasing pressure. Taking account of these phenomena, further increases in the film thickness resulted in uniform and dense films, as shown in figure 4. In this case, the film thickness is about 45 μ m. The crystal grain size in the film becomes larger. It was confirmed that there is no gas leakage in the thick films, which is important for its practical use as an electrolyte of SOFC.

As-deposited film had an amorphous structure. Figure 5 shows the X-ray diffraction pattern of the film after post annealing at 1273K for 10 hours. In this case, the LSGM films deposited on the NiO had the basic structure of LaGaO$_3$. Other crystalline structures such as La$_2$O$_3$, La$_4$Ga$_2$O$_9$ phases were also observed in some other films depending on the preparation condition. Similar XRD pattern was obtained for LSGMC films.

Figure 6 shows the dependence of electrical conductivity of LSGMC films on the temperature. The electrical conductivity of bulk LSGMC is also shown in the figure for

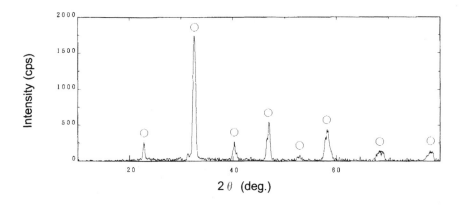

Figure 5. XRD pattern of the LSGM film after post annealing at 1273K for 10 hours. (○ : LaGaO$_3$ perovskite pattern)

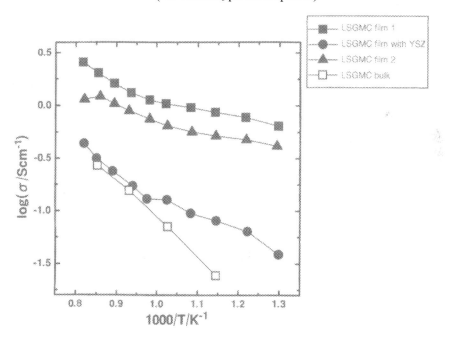

Figure 6. Arrhenius plot of electrical conductivity of LSGMC films. LSGMC films 1 and 2 were deposited on NiO substrate at the different target-substrate distances of 75mm and 85mm, respectively. LSGMC film with YSZ indicates the film deposited on the YSZ/NiO.

Table I. Compositions of the film deposited by PLD and of the bulk target.

	La	Sr	Ga	Mg	Thickness
Target	0.90	0.10	0.80	0.20	2mm
Film	0.86	0.128	0.78	0.148	17.6 μ m

comparison.   The LSGMC film did not show ionic conductivity.   This is due to Ni diffusion into the films during the annealing, causing electronic conduction.   With the preparation of buffer layer (YSZ) between the LSGMC film and NiO substrate, the electronic conduction was somewhat suppressed.

The composition of the annealed LSGM films was determined using SIMS.   Table I shows the compositions of the annealed film and the stoichiometric target.   A slight difference in stoichiometry between the bulk material and the film exists.   The atomic composition ratio of the films needs to be improved by optimizing the deposition conditions.

CONCLUSIONS

In this study, doped $LaGaO_3$ films were prepared on NiO substrates by pulsed laser deposition technique.   The following results were obtained:
(1)   The surface morphology of the films deposited on NiO substrate has remarkable dependence on ambient pressure.   Increasing the background pressure decreases both   the plume velocity and the deposition rate.
(2)   The films with a thickness greater than several 10 μ m showed an uniform and dense morphology.   Crystalline LSGM or LSGMC films could be deposited on the NiO substrates.
(3)   The film deposited on NiO substrate showed electronic conduction.   The electronic conduction could be suppressed by using a YSZ buffer layer.   The stoichiometry of the deposited films was slightly deviated from that of the target.

REFERENCES

1. T. Ishihara, H. Matsuda, and Y. Takita, *J. Am. Chem. Soc.*, **116**, 3801 (1994).
2. K. Kuroda, I. Hashimoto, K. Adachi, J. Akikusa, Y. Tamou, N. Komasa, T. Ishihara, Y. Takita, *Sol. State Ionics*, **132**, 199 (2000).
3. K. L. Saenger, "Pulsed Laser Deposition of Thin Films", ed. D. B. Chrisey and G. K. Hubler, (John Wiley & Sons, Inc., New York, 1994) pp.581-604.
4. T. Mathews, J. R. Sellar, and B. C. Muddle, *Chem. Mater.*, **12**, 917 (2000).

Mat. Res. Soc. Symp. Proc. Vol. 730 © 2002 Materials Research Society   V5.22

Single-Source Approach for The Growth of I-III-VI Thin Films

Mohammad Afzaal,[a] Theivanayagam C. Deivaraj,[b] Paul O'Brien,[a] Jin-Ho Park,[a] and Jagadese J. Vittal[b]

[a] The Manchester Materials Science Centre and Department of Chemistry, University of Manchester, Oxford Road, Manchester, M13 9PL, UK.
[b] Dept of Chemistry, 3 Science Drive 3, National University of Singapore, Singapore.
E-mail: paul.obrien@man.ac.uk; chmjjv@nus.edu.sg; jin-ho.park@man.ac.uk

ABSTRACT

The ternary chalcopyrite semiconductors I-III-VI are currently used for photovoltaic solar cell applications. In this study, $AgIn_5S_8$ thin films were prepared from a series of single-source bimetalorganic precursors, e.g. $[(PPh_3)_2AgIn(SC\{O\}R)_4]$ (R = alkyl) by aerosol assisted chemical vapour deposition (AA-CVD). The compounds can be used as single-source precursors for the deposition of ternary compounds (I-III-VI) by one-pot reaction using CVD process and they are found to be air stable, which is favourable in comparison with metal alkyl compounds, which are found to be pyrophoric. The optimum growth temperature for the preparation of these films on glass and Si(100) substrates, was found to be above 350 °C in terms of crystallinity, although deposition occurred at low temperatures. The films have been investigated using XRD, SEM and EDS. SEM analysis shows that all films are microcrystalline but have different morphologies depending on the growth temperatures. XRD results show evidence of the crystalline nature of these films. The results of this comprehensive study are presented and discussed.

INTRODUCTION

Chemists have been interested in synthesizing effective molecular precursors for various metal chalcogenides[1-4] in the past two decades. However, the possibility of single-source precursors for silver indium sulfide materials has been unexplored. Both $AgInS_2$ and $AgIn_5S_8$ are known to be semiconducting.[5-8] $AgInS_2$ finds applications as linear and non-linear optical materials. The band gap of $AgIn_5S_8$ is 1.80 eV (300K) and has been identified as making it a suitable candidate for photovoltaic solar cell applications.[9] Various metal monothiocarboxylates have been used as single-source precursors for metal sulfides.[10-12] Hence we attempted to synthesize bimetallic examples of such compounds which could be used as precursors to deposit silver indium sulfides.

There are also some initial reports concerning the use of single-source organometallic precursors for the deposition of $CuInS_2$ films. Hepp and co-workers reported spray CVD of $CuInS_2$ films using a single-source precursor [13], $(Ph_3P)_2Cu(\mu-SEt)_2In(SEt)_2$, which was previously reported by Kanatzidis and co-workers[14]. In their study, highly oriented $CuInS_2$ was deposited on Si(111) substrates at 400 °C. Recently they also prepared a number of ternary single-source precursors, based on the $[\{ER_3\}_2Cu(YR')_2In(YR')_2]$ (E = P, As, Sb; Y = S, Se and R = alkyl, aryl). [15]

In this paper we report the growth of silver indium sulfide thin films from single-source precursors of the type $[(Ph_3P)_2AgIn(SCOR)_4]$ [R = Me(1), Ph(2)] by AACVD experiments.

EXPERIMENTAL DETAILS

Precursor Synthesis

[(Ph₃P)₂AgIn(SCOMe)₄] (1): Indium chloride tetrahydrate (0.10 g, 0.34 mmol) dissolved in 12 mL MeOH was added to a CH_2Cl_2 solution (25 mL) of {(PPh₃)₂AgCl}₂ (0.23 g, 0.17 mmol). To this solution was added NaSC{O}Me prepared *insitu* by mixing 97.5 μL (1.36 mmol) of CH₃{O}CSH and 0.031 g (1.36 mmol) of sodium in 15 mL of MeOH. The solvents were removed under a N_2 flow. The product was extracted into 20 mL of CH_2Cl_2, filtered immediately, layered with 40 mL of petroleum ether and allowed to stand at 5°C over night. The creamy precipitate formed was filtered and washed with petroleum ether and Et_2O and dried under vacuum. Yield: 0.37 g (76%). Anal. Calcd for $C_{44}H_{42}O_4S_4P_2AgIn$ (mol. wt. 1047.71): C, 50.44; H, 4.02; S, 12.24. Found: C, 50.52; H, 4.02; S, 12.79. ¹H NMR (CDCl₃): δ, ppm 2.24 (s, 12H, CH_3), 7.28- 7.41 (m, 30H, (C₆H_5)₃P). ¹³C NMR (CDCl₃): δ, ppm. For thioacetato ligand: 34.1 (CH₃C{O}S), 206.8 (MeC{O}S). For the PPh₃: 129.4 (C₃, ³J(P-C) = 9.8 Hz), 130.8 (C₄), 132.1 (C₁, ¹J(P-C) = 25.1 Hz), 134.1 (C₂, ²J(P-C) = 16.3 Hz). ³¹P NMR: δ, ppm. 8.03.

[(Ph₃P)₂AgIn(SCOPh)₄] (2): the compound was also obtained by a similar procedure to that described for **1** but PhC{O}SH was used instead of MeC{O}SH. Yield: 79%. Anal. Calcd for $C_{64}H_{50}O_4S_4P_2AgIn$ (mol. wt. 1296.0): C, 59.31; H, 3.89; S, 9.90. Found: C, 58.36; H, 3.80; S, 9.66. ¹³C NMR (CDCl₃): δ, ppm. For thiobenzoato ligand: 127.8 (C$_{2/6}$ or C$_{3/5}$), 129.2(C$_{2/6}$ or C$_{3/5}$), 132.8 (C₄), 138.0 (C₁). For the PPh₃: 128.7 (C₃, ³J(P-C) = 9.8 Hz), 129.9 (C₄), 132.3 (C₁, ¹J(P-C) = 22.9), 133.7 (C₂, ²J(P-C) = 16.3). ³¹P NMR: δ, ppm. 7.28(s).

Deposition of films and characterizations

Aerosol Assisted Chemical Vapour Deposition (AACVD): Approximately 0.25g of precursor was dissolved in 30ml Toluene (or THF) and the solution was transferred into the round-bottomed flask. Six glass substrates (1 × 2 cm) were placed inside the reactor tube. The carrier gas flow rate was controlled by Platon flow gauges. The solution in the flask is placed in a water bath above the piezoelectric modulator of a humidifier, where aerosol droplets are generated and transferred by the carrier gas into a hot-wall zone. Then both the solvent and the precursor evaporate and the precursor vapour reaches the heated substrate surface where thermally induced reactions and film deposition take place. This homemade aerosol-assisted chemical vapour deposition kit consists of a two-neck flask, a PIFCO ultrasonic humidifier (Model No. 1077) and a CARBOLITE furnace.

Film characterizations: X-ray diffraction studies were performed using Cu-K_α radiation on a Philips X'Pert MPD diffractometer. The sample was mounted flat and scanned from 2θ 20 - 80 ° in steps of 0.04 ° with a count time of 2 s. Samples were carbon coated before electron microscopic analysis. A Jeol Superprobe 733 microscope was employed for all microscopy and EDS analyses.

RESULTS AND DISCUSSION

The residual weights observed for **1** and **2** in thermogravimetric analyses indicate the formation of $AgInS_2$. The decomposition occurs in unresolved multiple steps in the temperature range 132 - 315°C for **1** and 175 - 328°C for **2**. X-ray powder diffraction (XRD) of the final residue obtained from pyrolysis of compounds **1** and **2** indicated the formation of orthorhombic $AgInS_2$, JCPDS: 25-1328 (Fig 1). I-III-VI$_2$ materials are known to preferably adopt tetragonal structures, hence the formation of the high temperature orthorhombic $AgInS_2$ phase is interesting.

However, aerosol assisted chemical vapour deposition (AACVD) of THF solution of compounds **1** and **2** yielded $AgIn_5S_8$ films. Films grown on glass at 350, 400 and 450 °C from compound **1** were found to be transparent and adherent (Scotch tape test) with a dark red colour and films grown at 350 °C were slightly red.

XRD analysis (step size; 0.04°/2 sec) confirmed that the films prepared from compounds **1** and **2** were cubic-$AgIn_5S_8$ (JCPDS 25-1329) with a preferred orientation along the (311) plane (Fig 2). XRD patterns of films grown from compound **1** show the (111) and (222) planes were noticeably enhanced with increasing growth temperature and intensities of peaks from the (511) and (440) planes were also reversed. Measurements of films from compound **2** by XRD indicate similar patterns but no indication of enhancement of peak intensities was observed in the case of the films grown from compound **1**.

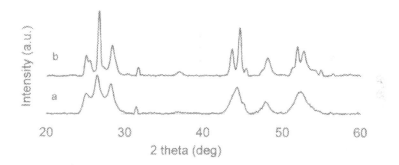

Fig. 1. X-ray powder diffraction of $AgInS_2$ formed by pyrolysis of compounds (a) **1** and (b) **2**.

Scanning electron microscopy (SEM) studies show there is a dramatic change in the morphologies of films grown from compound **1** at different growth temperatures (350 – 450 °C). The morphology of the films grown on glass at 450 °C (Fig. 3c) consists of thin plate-like particles, perpendicularly laid down onto the substrate with random orientation and similar morphology was also found in the films on Si(100) substrates grown at the same growth temperature. However, with decreasing growth temperature from 450 to 400°C, the morphology was dense and no longer plate-like particles were formed (Fig 4b). Growth rate and average particle size of the films grown at 400°C were *ca.* 0.3 μmh^{-1} and *ca.* 0.5 μm and at 350 °C *ca.* 0.2 μmh^{-1} and *ca.* 0.3 μm.

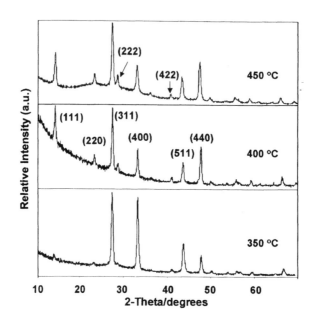

Fig 2. X-ray powder diffraction of $AgIn_5S_8$ thin films formed from compound **1** (temperatures indicate growth temperatures).

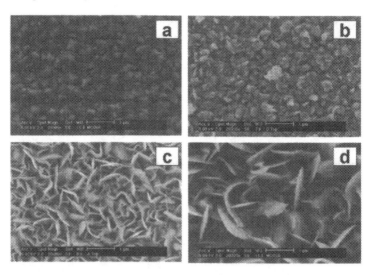

Fig 3. SEM images of $AgIn_5S_8$ films on glass grown from compound **1** (a: 350 °C on glass, b: 400 °C on glass, c: 450 °C on glass and d: 450 °C on Si(100)).

Morphologies of films grown from compound **2** (Fig. 4) are quite different from those of compound **1**. Films obtained at growth temp. of 350 °C revealed relatively dense morphology and a smaller particle size than those prepared from higher growth temperatures (*ca.* 100 – 300 nm). Particles on glass substrates (Fig. 4c) grown at 450 °C have trigonal habits consisting of several layers. However, on changing the substrate from glass to Si(100), the formation of particles leads to rice-like instead of triangular shapes (Fig. 4d). Growth rates obtained from cross-section views for the films were found to be *ca.* 0.2 μm/h at 350 °C, 0.5 μm/h at 400 °C and 0.6 μm/h at 450 °C on glass substrates after 2 hours growth.

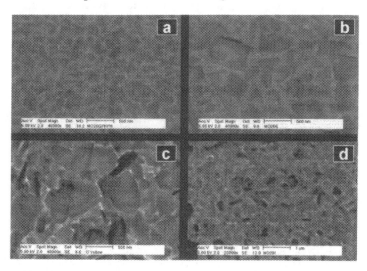

Fig 4. SEM images of AgIn$_5$S$_8$ films on glass grown from compound **2** (a: 350 °C on glass, b: 400 °C on glass, c: 450 °C on glass and d: 450 °C on Si(100).

CONCLUSIONS

In conclusion, compounds [(Ph$_3$P)$_2$Ag(μ-SC{O}R-*S*)$_2$In(SC{O}R)$_2$] (R = Me, Ph) are shown to be suitable single-source precursors for the deposition of cubic AgIn$_5$S$_8$ films on glass substrates by AACVD at 350, 400 and 450°C. The films grown at 350 °C and 400°C were denser on glass than those grown at 450 °C. Similar results have also been obtained for compound **2**.

We have demonstrated their potential use as single source precursors to silver indium sulfide thin films. Further studies of novel single-source precursors for the growth of I/III/VI thin films are in hand.

ACKNOWLEDGEMENTS

POB acknowledges the support of Sumitomo/STS as visiting Professor of Materials Chemistry at Imperial College, London, UK. Authors thank the EPSRC, UK and National University of Singapore for the grants that have made this research possible.

REFERENCES

1. M. Bochmann, *Chem. Vap. Deposition,* **2**, 85 (1996).
2. P. O'Brien and R. Nomura, *J. Mater. Chem.* **5**, 1761 (1995).
3. A.N. Gleizes, *Chem. Vap. Deposition,* **6**, 155 (2000).
4. M. Lazell, P. O'Brien, D.J. Otway and J.-H. Park, *J. Chem. Soc., Dalton Trans.* 4479 (2000).
5. S.C. Abrahams and J.L. Bernstein, *J. Chem. Phys.* **59**, 1625 (1973).
6. I. Yonenaga, K. Sumino, E. Niwa and K. Masumoto, *J. Cryst. Growth* **167**, 616 (1996).
7. A. Usujima, S. Takeuchi, S. Endo and T. Irie, *Jpn. J. Appl. Phys., Part 2* **20**, L505 (1981).
8. Y. Ueno, Y. Hattori, M.Ito, T. Sugiura and H. Minoura, *Sol. Energy Mater. Sol. Cells* **26**, 229 (1992).
9. N.M. Gasanly, A. Serpengüzel, A. Aydinli, O. Gürlü and I Yilmaz, *J. Appl. Phys.* **85**, 3198 (1999).
10. G. Shang, M.J. Hampden-Smith and E.N. Duesler, *Chem. Commun.* 1733 (1996).
11. M. D. Nyman, M. J. Hampden-Smith and E. N. Duesler, *Inorg. Chem.* **36**, 2218 (1997).
12. M. D. Nyman, K. Jenkins, M.J. Hampden-Smith, T.T. Kodas, E.N. Duesler, A.L. Rheingold and M.L. Liable-Sands, *Chem. Mater.* **10**, 914 (1998).
13. J. A. Hollingsworth, A. F. Hepp and W. E. Buhro, *Chem. Vap. Deposition* **5**, 105 (1999).
14. W. Hirpo, S. Dhingra, A. C. Sutorik and M. G. Kanatzidis, *J. Am. Chem. Soc.* **115**, 1597 (1993).
15. K. K. Banger, J. D. Harris, J. E. Cowen and A. F. Hepp, *Thin Solid Films* **403-404**, 390 (2002).

Materials for Power
in Space

Mat. Res. Soc. Symp. Proc. Vol. 730 © 2002 Materials Research Society V6.2

Ultralight Amorphous Silicon Alloy Photovoltaic Modules For Space Applications

Kevin Beernink, Ginger Pietka, Jeff Noch[1], Kais Younan, David Wolf, Arindam Banerjee,
Jeffrey Yang, Scott Jones[1], and Subhendu Guha
United Solar Systems Corp., 1100 W. Maple Rd.,
Troy, MI 48084, U.S.A.
[1]Energy Conversion Devices, Inc., 2956 Waterview Drive,
Rochester Hills, Michigan 48309, U.S.A.

ABSTRACT

Results for solar cells with high specific power (W/kg) using amorphous silicon alloy technology are reported. Currently available roll-to-roll production technology capable of high volume has been used to form cells on thinned stainless steel with high specific power. Results of cells on thinned stainless steel formed in batch mode in research and development (R&D) machines are also presented. Cells on polyimide in the early R&D stage have also been formed and are described. An analysis of cell component material and dimension changes to increase specific power shows that specific power ~2000 W/kg is possible as a long-term goal. Cells with specific power ranging from 300 to 400 W/kg for production cells with 8.5 % beginning of life AM0 efficiency to over 1200 W/kg for R&D cells on polyimide substrates are presented.

INTRODUCTION

The availability of low-cost, lightweight and reliable photovoltaic modules is an important component in reducing the cost of satellites and spacecraft. In terms of the harsh space environment, amorphous silicon (a-Si) alloy cells have several advantages over other material technologies [1]. The deposition process is relatively simple, inexpensive, and applicable for lightweight, flexible substrates. The temperature coefficient has been found to be between –0.2 and –0.3 %/°C for high efficiency triple-junction a-Si alloy cells, which is superior to crystalline Si devices for high temperature operations. Additionally, the a-Si alloy cells are relatively insensitive to electron and proton bombardment. In particular, modules incorporating United Solar's a-Si alloy cells have been tested on the MIR space station for 19 months with only minimal degradation [2].

Here we report on a-Si alloy solar cells in various stages of development with high specific power (W/kg) for space applications. Currently available roll-to-roll production technology is used to form cells on a stainless steel substrate. The substrate is thinned from a 5 mil initial thickness to a 1 mil final thickness to reduce the mass of the cells. In an effort to improve the specific power through higher efficiency, cells have been formed in batch mode in R&D machines, with the eventual goal of transferring these results to the production environment. A longer-term effort to develop cells on ultralight polyimide substrates is also underway, and cells have been formed in the R&D mode with very high specific power. In each case, the cells are optimized for the AM0 spectrum. An analysis of changes in cell component material and dimensions to increase specific power shows that it is possible, through a combination of efficiency improvement and mass reduction, to form large-area cells using a-Si alloy technology with specific power near 2000 W/kg.

CELL STRUCTURE AND RESULTS

It is well recognized that a multi-bandgap, multijunction cell structure offers the maximum advantage to obtain high efficiency in a-Si alloy solar cells and modules [1, 3]. The cells in this work use a triple-junction, spectrum splitting structure [3]. A schematic view of the cross-section of a cell is shown in Fig. 1. The Ag/ZnO back reflector layer is first deposited by sputtering onto either a stainless steel or polyimide substrate. The deposition parameters are such that the back reflector is textured to facilitate scattering and multiple reflections. The a-Si alloy layers are next deposited by radio-frequency plasma-enhanced chemical vapor deposition. A layer of Indium Tin Oxide (ITO) is deposited by sputtering, and serves as a transparent conductor and anti-reflective coating. The cells are next passivated to remove any shunts and shorts existing in the large area cells. Wire grids and bus bars are added to the top of the structure. For cells on stainless steel substrates, a bus bar is also added to the back of the cell, opposite to the top bus bars. A top view of a cell showing the layout of the wires and bus bars is shown in Fig. 2. The bus bars at either side serve as the positive cell contacts. For cells on polyimide, a portion of the substrate coated with back reflector extends out past the bus bars to serve as the negative cell contact. For cells on stainless steel, the substrate is cut just outside the bus bars, and the bus bars on the substrate are the negative contacts. Standard dimensions are given in Table I.

The specific power of the cells can be increased either by increasing efficiency or reducing mass, or both. Various enhancements to the materials in the a-Si alloy layers are utilized to increase the efficiency of the cells. The a-Si:H i layer of the top cell is deposited so that it is near the transition to microcrystallinity. This has been shown to give cells with the highest efficiency [4]. The p layers are microcrystalline to reduce absorption and to allow for higher electrical conductivity, thus increasing the open-circuit voltage [5]. For the middle and bottom cells, the ratio of Si to Ge is graded to add a profile to the bandgap to improve carrier collection [6]. In order to have the highest fill factor, the top cell of the triple-junction structure should be the current-limiting component cell. The cells have been optimized for the AM0 spectrum by adjusting the thicknesses and compositions of the i layers in order to achieve proper current matching of the three component cells.

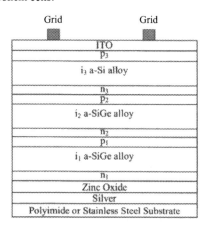

Figure 1. Cross-section of the triple-junction a-Si alloy solar cell.

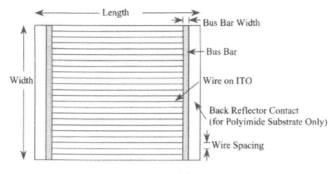

Figure 2. Top view of a cell. For cells on polyimide substrates, a portion of the substrate coated with back reflector extends past the bus bars to form the negative cell contact.

Table I. Standard cell components and dimensions.

Wire Spacing	Top Bus Bar			Back Bus Bar (stainless steel substrates)		
	Material	Width	Thickness	Material	Width	Thickness
0.568 cm	Cu	0.55 cm	100 µm	Cu	0.635 cm	100 µm

Cells on thinned stainless steel substrates

Cells with the structure described above have been formed on stainless steel substrates using present roll-to-roll production technology, as well as in batch mode in R&D machines. Table II lists substrate dimensions, aperture area efficiency and specific power under AM0 conditions for two sizes of cells made with the present production technology and for present R&D cells. The present production and present R&D cells have standard components as in Table I. All of the cells on stainless steel substrates are initially formed on 5 mil stainless steel and subsequently undergo thinning of the substrate to a final thickness of ~0.75 mil. The thinning process does not affect the efficiency of the cells. As seen in Table II, cells have been manufactured using present production technology with specific power as high as 396 W/kg.

Table II. Present and projected specific power for cells on thinned stainless steel substrates.

		Substrate Dimensions	Aperture Area (cm^2)	AM0 Eff (%)	AM0 Specific Power (W/kg)
Present Production Technology	H-strip, Cu top and back buses	35.56 cm x 4.0 cm x 0.75 mil	131	8.42	396
	L-strip, Cu top and back buses	35.56 cm x 23.9 cm x 0.75 mil	809	8.42	384
Projected Production	L-strip, Al top bus, small Al tabs on back	35.56 cm x 23.9 cm x 0.50 mil	809	10.7	961
Present R&D	460 cm^2, Cu top and back buses, NASA measurement	22.8 cm x 21.6 cm x 0.75 mil	460	9.49	355
Projected R&D	460 cm2, Al top bus, small Al tabs on back	22.8 cm x 21.6 cm x 0.50 mil	460	10.9	885

Although the efficiency of the present R&D cells is significantly higher than the production cells, this does not show up in the specific power. This is a result of the different aspect ratio of the cells. The ratio of bus bar mass to substrate mass is larger for the R&D cells.

Also shown in Table II is a projection of results expected to be achieved shortly in R&D through improvement in efficiency and mass reduction. Efficiency gains are expected through further component cell optimization and current matching in the triple-junction structure. For the projection, an aperture area efficiency of 11.0 % was used for a cell with the standard Cu top and back buses, as currently used for the production and R&D cells. Then, an analysis of cell component material and dimensional changes to reduce the mass while accounting for increased electrical losses was used to maximize specific power. No changes are made to the semiconductor layers once efficiency has been optimized. The analysis is described in detail below. The optimized cell is on a 0.5-mil-thick substrate, uses 10-μm-thick Al top bus bars in place of the standard Cu, has small Al tabs instead of Cu bus bars on the back, and the wire spacing is increased to 0.95 cm. The cell has been optimized for specific power, rather than efficiency, and the 10.9 % efficiency is lower than the 11.0 % of the reference (heavier) case, because some of the changes for reduced mass have associated electrical losses. Of the 149 % increase in specific power for the optimized case compared to the present R&D case, only 16 % comes from an increase in efficiency. The remainder of the gain comes from changes that reduce the mass of the cell. A similar projection is shown for the L-strip production cell. The projection is done as for the R&D cell and uses the same parameters, except that the optimized bus bar thickness is 12.5 μm. Again, because of the difference in aspect ratio, the projection for the production-sized cell has significantly higher specific power than the projected R&D cell, even though the assumed efficiency is the same.

Cells on polyimide substrates

As a longer-term alternative to the presently available or soon-to-be-improved production cells on thinned stainless steel substrates, ultralight polyimide substrates are being used in R&D to make cells with extremely high specific power. Triple-junction cells with the structure shown in Figs. 1 and 2 have been formed on 1-mil-thick polyimide substrates. Standard components and dimensions were used, as in Table I, except that the top bus bars are Al, and there are no back bus bars. Instead, the extensions coated with back reflector, as in Fig. 2, are used for the negative contact. Table III shows the results for the present R&D structure as measured at the NASA Glenn Research Center under AM0 conditions using a LAPSS solar simulator. The output power is 5.05 W with a mass of 4.02 g and specific power of 1256 W/kg. Although this work is in its early stages, this result nevertheless represents a world record specific power for a realized cell or module with output power of this magnitude.

Table III also gives a projection of what is possible with efficiency improvement and a lighter structure. As for the stainless steel projection, an analysis of cell components was done to maximize specific power. An aperture area efficiency of 11.0 % was assumed for a cell with the standard Cu top bus bars, as in Table I. Then, changes to the cell components were analyzed as described below to find the configuration for maximum specific power. A wire spacing of 1.2 cm and a very thin, 5-μm-thick Al top bus bar result in a specific power projection of over 2000 W/kg. As is the case for stainless steel, there is much room for improvement in specific power even if the assumed efficiency is not reached.

Table III. Present and projected specific power for cells on polyimide substrates.

		Substrate Dimensions	Aperture Area (cm2)	AM0 Eff (%)	AM0 Specific Power (W/kg)
Present R&D	Polyimide, Al top bus (NASA measurement)	25.1 cm x 20.5 cm x 1 mil	412	9.0	1256
Projected R&D	Polyimide, Thin Al top bus, 1.2 cm wire spacing	22.4 cm x 20.5 cm x 1 mil	412	10.5	2070

Specific power optimization

An analysis was undertaken to investigate how cell component materials and dimensions could be changed to increase specific power. The analysis considers the mass savings and corresponding power losses introduced by changes in some cell components. Changes considered are 1) substrate thickness, 2) wire spacing, and 3) top bus bar material and dimensions. Electrical losses are calculated as the I^2R losses for current flow through the ITO to the wires, through the wires to the bus bars, and through the bus bars to the center of the bus bars, where connections to the cell are made on each side. The current used in the electrical loss calculations is the expected current at maximum power for AM0 operation for the given efficiency. For the I^2R loss calculation, distributed current generation was taken into account, and results in a multiplying factor of 1/3 for the I^2R loss compared to the case of constant current flow along the length of a component. For changes in the wire spacing, the change in shadow loss is also included.

As expected, a reduction in substrate mass has a major effect on specific power. For the bus bars, mass reduction can be achieved by using Al in place of Cu, since only half the mass is necessary to achieve the same resistance. The thickness of the bus bar can be reduced for further mass reduction, but with an increase in resistance and associated electrical loss. A reduction in wire mass is considered by increasing the spacing between wires. The reduced mass is accompanied by reduced shadowing, but is offset somewhat by a higher series resistance and associated I^2R loss for current flow through the ITO to the wires. The standard 0.568 cm wire spacing used in the present cells was chosen based on maximum efficiency, and is determined by the trade-off between shading and electrical loss in the wires. Calculation shows that for all but the lightest cells (1 mil polyimide or 0.5 mil stainless steel), very little increase in specific power can be gained by a change in wire spacing.

Figure 3 shows the results of a calculation of specific power as a function of bus bar thickness for the present and projected production L-strip, present and projected R&D cells on stainless steel, and present and projected R&D cells on polyimide, corresponding to cells in Tables II and III. The present production L-strip and present R&D cell have the standard wire spacing and bus bar dimensions as in Table I, except for the variation in top bus bar thickness. The substrate thickness is 0.75 mil. The projected production L-strip and projected R&D stainless steel cells use a 0.5 mil substrate, 0.95 cm wire spacing, Al top bus bars, and small Al tabs for the back contact, as described above. The present polyimide cell also uses the standard wire spacing and top bus bar dimensions, but the top bus bar material is Al. The projected polyimide cell has Al top bus bars and 1.2 cm wire spacing. Both polyimide cells are on a 1 mil substrate. It can be seen in Fig. 3 that specific power up to 961 W/kg and 2070 W/kg are projected for the stainless steel and polyimide cells, respectively.

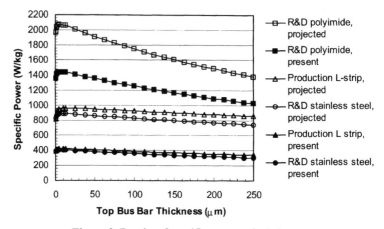

Figure 3. Results of specific power calculation.

CONCLUSIONS

Solar cells using a-Si alloy technology on stainless steel and polyimide substrates with high specific power have been formed. Using present production technology, cells on thinned stainless steel have specific power as high as 396 W/kg. Projected production cells on stainless steel with expected efficiency gains and mass reduction have specific power of 961 W/kg. R&D cells on 1 mil polyimide with specific power of 1256 W/kg have been achieved, and specific power up to ~2000 W/kg is projected for improved cells.

ACKNOWLEDGEMENTS

The authors would like to thank Tufail Nazmee, Blong Hang, and Eric Akkashian for assistance in cell fabrication, David Scheiman and Phillip Jenkins for cell measurements at NASA Glenn Research Center, and Stan Ovshinsky for stimulating discussions. This work was funded in part by the U.S. Air Force under contract F29601-00-C-0024.

REFERENCES

1. S. Guha, J. Yang, A. Banerjee, T. Glatfelter, G.J. Vendura, Jr., A. Garcia, and M. Kruer, *Proc. 2nd World Conf. on Photovoltaic Solar Energy Conversion, Vienna*, 3609 (1998).
2. M. Kagan, V. Nadorov, S. Guha, J. Yang, and A. Banerjee, *Proc. 28th IEEE PV Specialists Conf., Anchorage*, 1261 (2000).
3. J. Yang, A. Banerjee, and S. Guha, *Appl. Phys. Lett.* **70**, 2975 (1997).
4. J. Yang and S. Guha, *Mater. Res. Soc. Proc.* **557**, p. 239 (1999).
5. S. Guha, J. Yang, P. Nath, and M. Hack, *Appl. Phys. Lett.* **49**, 218 (1986).
6. S. Guha, J. Yang, A. Pawlikiewicz, T. Glatfelter, R. Ross, and S. R. Ovshinsky, *Appl. Phys. Lett.* **54**, 2330 (1989).

Disordered and Nanoscale Materials
for Energy Applications

Mat. Res. Soc. Symp. Proc. Vol. 730 © 2002 Materials Research Society

Carbon Nanotube-Perovskite-Composites as New Electrode Material

Anke Weidenkaff[1], Stefan G. Ebbinghaus[1], Thomas Lippert[2], Macarena J. Montenegro[2], and Armin Reller[1]
[1]Solid State Chemistry, University of Augsburg, D-86159 Augsburg, Germany,
[2]Dept. General Energy Research, Paul Scherrer Institute, CH-5232 Villigen, Switzerland

ABSTRACT

In this paper we describe the synthesis and characterisation of $La_{1-x}A_xCoO_3$ (A= Ca, Sr) ($0 < x < 0,5$) of different morphologies using pulsed laser deposition, ceramic methods and alternative soft-chemistry techniques. Furthermore the potential use of a $La_{1-x}Ca_xCoO_3$ /carbon nanotube composite material for oxygen electrodes is discussed.

INTRODUCTION

The stability and efficiency of oxygen electrodes of metal - air batteries and alkaline fuel cells will be decisive for their technical realization. The redox processes at the electrode in the highly alkaline media are responsible for the short lifetime of conventional teflon-bonded carbon diffusion electrodes. The oxygen evolution and oxygen reduction reactions in alkaline electrolytes occur via a two- electron pathway involving the formation of intermediate peroxide ions and OH^- [1].

Perovskite type metal oxides (ABO_3) with a lanthanide ion in the A position and a transition metal ion in the B position are known to be low cost, stable and active electrocatalysts for practical applications in fuel cells and metal-air batteries [2,3,4]. Particularly lanthanum cobaltates, manganates, ferrates and nickelates show good catalytic properties for the peroxide decomposition [5] [6].

For the production of gas diffusion electrodes, calcium and strontium substituted rare earth cobaltate powders with perovskite structure are synthesised. To study the influence of the particles size and shape co-precipitation-, ceramic-, complexation- and microemulsion processes are applied. As perovskites are not highly conductive, high surface area carbon and teflon as binder has to be added for a functional composite electrode material. We prepared a carbon nanotube composite material by a catalytic hydrocarbon dissociation reaction on metal oxide particles. With the combination of the properties of the metal containing part and the carbon nanotubes in a carbon nanotube/metal oxide-composite the carbon/metal oxide interface as well as the chemical and thermal stability of the electrode material can be improved (as carbon nanotubes are known to be more stable than carbon black).

Since carbon itself is catalytically active for the oxygen reduction and evolution reaction as well, it is necessary to prepare additionally electrodes on inactive substrates and with well defined electrolyte/catalyst interfaces to study and compare the oxygen reduction/evolution reactions for different perovskites. In our experiments a modified pulsed laser deposition (PLD) technique with reactive gas pulses [7] is applied to grow dense crystalline La- cobaltate films on various substrates.

EXPERIMENTAL

Coarse powders of Ca- and Sr substituted La- cobaltates are produced at 1200°C with conventional solid state reactions. By mixing the components on the molecular level, the formation of the perovskite proceeds at lower temperatures. This can be achieved by the production of mixed metal nitrate crystals. Another way to mix the metal ions are co-precipitations in microemulsions [8]. By this method stable dispersions of aqueous droplets are serving as microreactors for the precipitation reaction. They are produced by stirring a mixture of oil phase, surfactants, and an aqueous phase. Two of these emulsions, one containing the metal nitrates and the other one the precipitation agent are mixed to form a homogeneous precipitate inside the micelles. In additional synthesis procedures metal complexes with organic ligands are prepared as precursors for the metal oxides. In the hydroxy carbon acid aided synthesis [9] [10] citric acid or tartaric acid is used as complexing agent. The precursors are calcined at temperatures between 600°C and 700°C.

The La, Ca – cobaltate films are grown on MgO and stainless steel substrates by PLD. The experimental set-up has been described previously in detail [11]. A rotating $La_{0.6}Ca_{0.4}CoO_3$ target is ablated and deposition takes place on a preheated substrate.

For the production of the carbon nanotube/metal oxide composite material, perovskite powders or macrostructured agglomerates (textile templates) are coated with a citrate precursor containing the perovskite mixture with a slight excess of cobalt oxide (< 1%). The material is produced in a similar procedure as described in [12].

The structural analysis of the perovskite powders and composite materials is done by powder X-ray diffraction (XRD) data collected on a Seifert XRD 3003-TT diffractometer using Cu-K_α radiation. Structure and texture of the single crystalline films have been studied on a Siemens D5000 X-ray diffractometer with Bragg-Brentano geometry (Cu K_α radiation). The apparatus is equipped with an Eulerian cradle for the sample orientation. X-ray pole figures were measured by rotating the sample around the φ axis and tilting the sample along the χ axis during the measurements with a fixed detector position (2θ).

The morphology of the samples was examined using a LEO Gemini 982 scanning electron microscope (SEM) equipped with a Röntec energy dispersive X-ray analysis (EDX) detector system. Transmission electron microscopic (TEM) studies and electron diffraction were performed on a Phillips CM 30 instrument equipped with EDX detector. The surface area is measured by the BET-method (Micrometrics ASAP 2000).

After preparation and characterisation of the perovskites, electrodes are produced and tested for their catalytic activity in an electrochemical treatment. The tests of the electrodes are performed during several charge and discharge processes [5] in a three electrode arrangement.

RESULTS AND DISCUSSION

The morphology of the perovskite particles depends strongly on the synthesis procedure. The products from the ceramic synthesis route are powders with a particle diameter of > 5 µm. The decomposition of the single crystalline mixed nitrates leads to well crystallised platelet-shaped particles as can be see on the electron micrograph (Fig. 1). The XRD and SAED pattern reveal a cubic cell with **a** in the range of 0,381 nm < a < 0,383 nm depending on the synthesis procedure and substitution on A-site. In the SAED pattern besides the cubic perovskite lattice reflections weak additional reflections are present indicative of a superstructural ordering [13] [14]. The EDX studies revealed (semiquantitatively) that the individual particles had the composition given by their chemical formula.

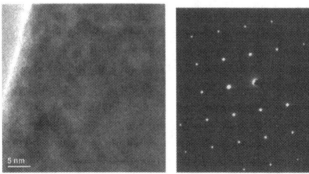

Fig. 1 : HRTEM and SAED od $La_{0.6}Ca_{0.4}CoO_3$ particles from the decomposition reaction of La, Ca, Co-nitrate single crystals

The XRD and TEM characterisation of the single crystalline $La_{0.6}Ca_{0.4}CoO_3$ films on MgO reveal that the oxide is growing epitaxially on the MgO substrate. The (200) and the (110) peaks of the product are the only non-substrate peaks appearing in the XRD Θ-2Θ scan at $\chi = 0$ (see Fig. 2a). Therefore it can be concluded that the $La_{0,6}Ca_{0,4}CoO_3$ is preferentially orientated along the (200) direction of the MgO.

Figure 2b shows the (111) pole figure of the $La_{0,6}Ca_{0,4}CoO_3$ -film at a 2θ angle at 42, 97°. In the figure four maxima at $\chi = 54°$ are visible, indicative of the single crystalline domains of the orientated film.

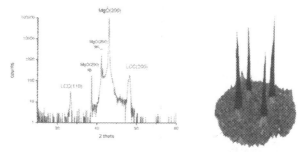

Fig. 2: XRD pattern (a) and 111 pole figure (b) of the $La_{0,6}Ca_{0,4}CoO_3$ film on MgO

TEM studies are performed to obtain detailed information on the local structure of the $La_{0,6}Ca_{0,4}CoO_3$ (LCC) films on MgO. The HRTEM image shown in Fig.3 represents the MgO substrate (upper part) and the $La_{0,6}Ca_{0,4}CoO_3$ -film (lower part).

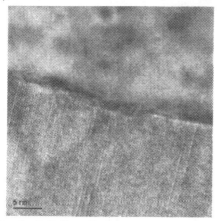

Fig. 3: HRTEM study of the $La_{0,6}Ca_{0,4}CoO_3$ film on MgO

The aim of this work is to optimise the electrode material by increasing its reactivity. Therefore the reactivity of the produced compounds is tested in two different processes: The catalysis of a hydrocarbon cracking reaction to form carbon nanotubes and the electrocatalytic performance in for oxygen evolution and reduction.

$La_{1-x}(Ca, Sr)_xCoO_3$ and the amorphous precursors of these compounds are used as catalysts for the growth of carbon nanotubes. The nanoclusters of cobalt formed on the surface of the perovskite particles during the nanotube growth are suitable nucleation sites for the growth of nanotubes. The reactivity towards hydrocarbon cracking for the applied cobaltates varies from a CNT growth rate of nearly 0 to a rate of 90 weight % per minute. The composition and texture of the catalyst decides upon product quality and yield. Nanoparticles from the amorphous precursor process or the precursors themselves produce more carbon nanotubes than coarse products from ceramic synthesis. Furthermore it was found that Ca substituted La- cobaltates are more active than Sr-substituted or pure

LaCoO₃ demonstrating that the average A-cation radius and the cation-size mismatch affects the catalytic properties of the cobaltates significantly. Ca- ions are smaller than La- and Sr ions. Therefore Ca-substituted La-cobaltates are probably less stable (Goldschmidt-factor) than their Sr-substituted analogues. To study the influence of the electronic and crystallographic structure on the catalytic reactions combined Rietveld and EXAFS investigations [15] are in progress.

The TEM and SEM images of the products reveal that the formed multiwalled carbon nanotubes (MWNTs) have a length of approximately 20 μm (see Fig.5) and diameters in the range of 20-30 nm. Some of the encapsulated particles were identified as metallic Co by EDX point analysis and EXAFS measurements. The data show clearly that during the reaction, the perovskite structure was partially destructed.

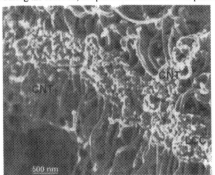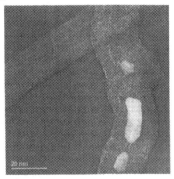

Fig. 5: SEM and TEM pictures of the LCC-CNT nanocomposite material

In additional experiments, perovskite particles are impregnated in advance with a precursor solution containing the appropriate stoichiometric citrate mixture with an excess of 5 % cobalt. The excess Co in the sacrifying layer forms the nucleation sites for the carbon nanotube formation and prevents the bulk perovskite phase from decomposition. With these precursors it is possible to grow carbon nanotubes only on the surface of the particles in a very short reaction time (3 min).

First tests with air electrodes consisting of 50 weight % multiwalled carbon nanotubes enclosing 50 weight % metal and metal oxide nanoparticles showed a comparable performance to state-of-the-art electrodes developed at PSI.

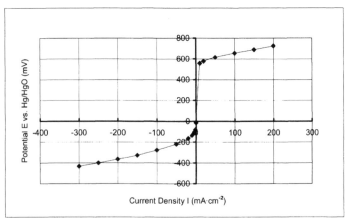

Fig. 6: Electrode Performance of the LCC-CNT material

Consequently these new composite materials are highly interesting candidates for technical applications in oxygen electrodes.

ACKNOWLEDGEMENTS

The authors would like to thank Stefan Müller, PSI for inspiring discussions, Roland Wessiken, ETH Zürich for his help in the transmission microscopic studies, Andreas Heinrich, Universität Augsburg for helping to do the XRD-measurements of the films, Frederike Geiger, PSI Villigen for the BET measurements, Gertraud Masanz, PSI for the production of the electrodes, and Franziska Holzer, PSI for the measurements on the electrodes.

REFERENCES

1. V. Hermann, D. Dutriat, S. Müller, and Ch. Comninellis, *Electrochimica Acta*, **46**, 365-372, (2000).

2. A. Martinez-Juarez, L. Sanchez, E. Chinarro, P. Recio, C. Pascual, and J. R. Jurado, *Solid State Ionics*, **135**, 525-528, (2000).

3. A. J. McEvoy, *J.Mater.Sci.*, **36**, 1087-1091, 2001.

4. Y. Ohno, S. Nagata, and H. Sato, *Solid State Ionics*, **9&10**, 1001-1008, (1983).

5. S. Müller, K. Striebel, and O. Haas, *Electrochimica Acta*, **39**, 1661-1668, (1994).

6. M. Bursell, M. Pirjamali, and Y. Kiros, *Electrochimica Acta*, **47**, 1651-1660, (2002).

7. M. J. Montenegro, T. Lippert, S. Müller, A. Weidenkaff, P. R. Willmott, and A. Wokaun, *Appl. Surf. Science*, in press (2002).

8. A. Weidenkaff, S. Ebbinghaus, and T. Lippert, *Chem.Mater.*, in press (2002).

9. M. S. G. Baythoun and F. R. Sale, *J.Mater.Sci.*, **17**, 2757-2769, (1982).

10. Y. Teraoka, H. Kakebayashi, I. Moriguchi, and S. Kagawa, *Chemistry Letters*, 673-676, (1991).

11. M. J. Montenegro, T. Lippert, S. Müller, A. Weidenkaff, P. R. Willmott, and A. Wokaun, *Phys.Chem.Chem.Phys.*, in press (2002).

12. A. Weidenkaff, S. Ebbinghaus, P. Mauron, A. Reller, Y. Zhang, and A. Züttel, *Materials Science and Engineering C*, **19**, 119-123, (2001).

13. A. Baiker, P. E. Marti, P. Keusch, E. Fritsch, and A. Reller, *J.Catal.*, **146**, 268-276, (1994).

14. R. H. E. van Doorn and A. J. Burggraaf, *Solid State Ionics*, **128**, 65-78, (2002).

15. S. G. Ebbinghaus, A. Weidenkaff, and R. J. Cava, *J.Solid State Chemistry*, in press (2002).

Mat. Res. Soc. Symp. Proc. Vol. 730 © 2002 Materials Research Society V7.3

Preparation and Characterization of Nanostructured FeS$_2$ and CoS$_2$ for High-Temperature Batteries

Ronald A. Guidotti[1], Frederick W. Reinhardt[1] Jinxiang Dai[2], and David E. Reisner[2]
[1]Sandia National Laboratories, P.O. Box 5800, Albuquerque, NM 87185-0614
[2]US Nanocorp®, Inc., 74 Batterson Park Rd., Farmington, CT 06032

ABSTRACT

In this paper, we report on the preparation of synthetic FeS$_2$ and CoS$_2$ using a relatively inexpensive aqueous process. This avoids the material and handling difficulties associated with a high-temperature approach. An aqueous approach also allows ready scale-up to a pilot-plant size facility. The FeS$_2$ and CoS$_2$ were characterized with respect to their physical and chemical properties. The synthetic disulfides were incorporated into catholyte mixes for testing in single cells and batteries over a range of temperatures. The results of these tests are presented and compared to the performance of natural FeS$_2$ (pyrite) and a commercial source of CoS$_2$.

INTRODUCTION

Conventional thermally activated ("thermal") batteries employ pyrite (FeS$_2$) as the primary cathode material with Li(Si) and Li(Al) alloys as anodes. These batteries are ideal for power sources for many military applications (e.g., missiles) because of their long shelf life. Similar technology is also being considered for potential domestic applications such as power supplies for geothermal borehole data logging.

Pyrite is a relatively abundant, inexpensive naturally occurring mineral. However, before it can be used for electrochemical applications, it must be ground, sized, and then purified to remove electroactive impurities and inert gangue material. The overall processing adds greatly to the cost of the final cathode material. Synthetic pyrite has a number of advantages over natural pyrite. This material is typically nanostructured, which results in a much higher surface area and a correspondingly higher current-carrying capability. It can be easily prepared in high purity in large scale from inexpensive precursors [1-4]. This avoids much of the nonhomogeneity associated with natural pyrite from various sources. The Co analogue to pyrite, CoS$_2$, is preferred to pyrite for applications involving high current densities, due to its much higher electronic conductivity and thermal stability (650° vs. 550°C for FeS$_2$). It is available commercially in kg quantities from Cerac, Inc. (Butler, WI). The particle size of the natural pyrite can be controlled by selective grinding. However, no particle-size control is presently possible with the commercial CoS$_2$.

Sandia has extensively characterized such disulfides in many thermal batteries over the years and has a large database as a reference point. We studied an aqueous route for the preparation of synthetic FeS$_2$ and CoS$_2$ with the goal of developing a viable large-scale (e.g., 1-kg batches) process that consistently produces quality material suitable for high-temperature batteries. Some control over product particle size was also desired. This report documents the results of that study.

EXPERIMENTAL

Disulfide Preparation

The metal disulfides were prepared at US Nanocorp (USN) by a synthesis process in which an aqueous solutions of either 1.9M $CoCl_2 \cdot 6H_2O$ or 1.3M $FeCl_3$ were added to a second aqueous solution of Na_2S_2 containing the required stoichiometric amount of polysulfide. Both solutions were heated to 80°C before mixing and were maintained at this temperature for 11 h and then allowed to cool to ambient. After vacuum filtering and washing, the filter cake was heat treated at temperatures of 200° to 400°C for FeS_2 and 400° to 550°C for CoS_2. All preparation, reaction, filtering, and heat treatment were carried out under a nitrogen cover and the product was stored under dry nitrogen because of the high reactivity of the resultant disulfide product with ambient air. Batches of up to 500 g were made. The reaction in the case of CoS_2 is shown in equation 1.

$$CoCl_{2(aq)} + Na_2S_{2(aq)} \xrightarrow{\hspace{1cm}} CoS_{2(s)} + 2NaCl_{(aq)} \qquad [1]$$

Electrochemical Testing

Flooded anodes (0.9 g) of 25% LiBr-KBr-LiCl eutectic electrolyte and 75% active anode (44% Li/56% Si, Foote) were used for all tests. The separator pellet (1.0 g) was formulated with 35% Maglite 'S' MgO (Merck) and LiBr-KBr-LiCl eutectic, which melts at 321°C. A catholyte was formulated with 73.5% disulfide, 25% separator, and 1.5% Li_2O as a lithiation source and was fused under argon at either 450°C (for FeS_2) or 500°C (for CoS_2) for 4 h. Natural pyrite (-325 mesh, American Minerals, HCl purified) and commercial CoS_2 (Cerac, Inc., Butler WI) were used as control materials. All processing of powders and materials and cell assembly was conducted in a dry room maintained at <3% relative humidity.

The single cells were 1.25" (3.18 cm) in diameter and were tested galvanostatically (constant current) under computer control in a glovebox in high-purity argon that contained <1 ppm each water and oxygen. The single cells were tested under computer control using a programmed galvanostat (PAR 371). A load of 1.0 A (125 mA/cm^2) was used as a background with 2.0-A pulses (500 ms to 1 s long) being applied every 30 s. The cells were tested at a temperature of 400° and 500°C (to bracket the normal operating range of a thermal battery) to a cutoff voltage of 1.0 V. Limited 5-cell battery tests were also performed under similar discharge conditions using a reusable test fixture with provisions for measurement of the stack temperature.

RESULTS AND DISCUSSION

Physical Properties

X-ray diffraction (XRD) analysis was performed on the synthesized disulfides. The materials were amorphous as prepared but were crystalline after thermal treatment. The synthetic FeS_2 heat treated at 400°C showed only lines of pyrite (FeS_2), while the synthetic CoS_2 showed cattierite (CoS_2) to be the major phase with linnaeite (Co_4S_3) as a minor phase when heat treated at 400° or 500°C. After heating at 550°C, the only major phase was cattierite. The average grain sizes estimated from half-height width analysis of the spectra were 28 – 39 nm and 40 – 45 nm for the 400° and 500°C CoS_2, respectively. The average particle size of the natural pyrite was ~44 μm while that of the Cerac CoS_2 was 64 nm. The sample morphologies of the synthetic FeS_2 and

CoS$_2$ are shown in Figures 1a and 1b, respectively. As can be seen, the CoS$_2$ particles are much smaller than the FeS$_2$ particles, which explains its ready ease of oxidation by ambient air.

During thermal decomposition, FeS and elemental sulfur are formed in the case of FeS$_2$ while Co$_3$S$_4$ and sulfur are formed in the case of CoS$_2$. The thermal stabilities of the synthetic sulfides are shown in the thermogravimetric analysis (TGA) data of Figures 2a and 2b, respectively, for FeS$_2$ and CoS$_2$. Comparable data for natural pyrite and a commercial CoS$_2$ are shown for comparison. The synthetic FeS$_2$ showed a similar thermal stability as the natural pyrite. When heat treated at 400° and 500°C, the synthetic CoS$_2$ showed a lower stability than the commercial material. After heat treatment at 550°C, however, the thermal stability improved and was comparable to that of the commercial CoS$_2$. Thus, thermal treatment can have a significant impact on the structure and thermal properties of certain synthetic metal disulfides.

Electrochemical Properties

Single-Cell Tests – The performance of Li(Si)/FeS$_2$ single cells at 125 mA/cm^2 at 400° and 500°C are shown in Figures 3a and 3b, respectively for two lots of synthetic FeS$_2$. At 500°C (Figure 3b), the performance of the synthetic pyrites was comparable to that of the natural pyrite. Although the overall polarization was somewhat greater for this material at the start of discharge for the cells with synthetic pyrite, the polarization remained relatively flat and was less later in life. However, at 400°C (Figure 3a), the initial performance of the cells with the synthetic pyrite was comparable to that of the natural pyrite until the first voltage transition where the polarization rapidly increased, degrading performance. This may be due to inadequate wetting of the nanostructured FeS$_2$ by electrolyte and the development of a higher-impedance interface between the first discharge phase, Li$_3$Fe$_2$S$_4$, and the separator. This is generally not a problem with the relatively coarse natural pyrite particles. Tests are underway in which the nanostructured pyrite will be pretreated to enhance wetting by electrolyte to improve the electrochemical performance.

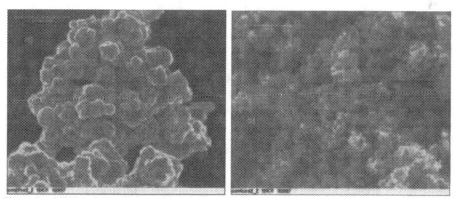

Figure 1. a) SEM photomicrograph of synthetic FeS$_2$ (1.5 μm marker) heated to 400°C and b) synthetic CoS$_2$ (500 nm marker) heated to 500°C.

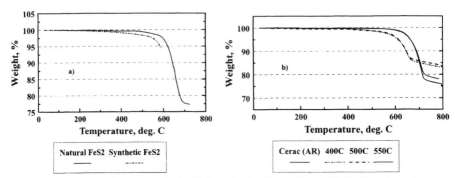

Figure 2. Thermogravimetric analysis of a) synthetic FeS_2 and b) synthetic CoS_2 heated under argon at 10°C/min.

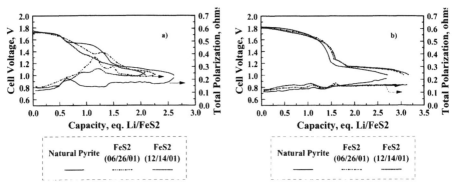

Figure 3. Performance of Synthetic FeS_2 in Li(Si)/LiBr-KBr-LiCl (MgO)/FeS_2 Cells at 125 mA/cm^2 at a) 400°C and b) 500°C.

The performance of Li(Si)/CoS_2 single cells at 125 mA/cm^2 at 400° and 500°C are shown in Figures 4a and 4b, respectively for Cerac CoS_2 and USN CoS_2 heat treated at various temperatures. The heat treatment used for the USN materials had a dramatic affect on the performance at 400°C (Figure 4a). Treatment temperatures of 400° and 500°C resulted in material with much higher impedance and reduced life. A voltage transition at ~0.7 eq. Li/CoS_2 occurred for these materials but was not observed for the material heat treated at 550°C or for the Cerac CoS_2. Even at the highest heat-treatment temperature of 550°C, the performance of the USN CoS_2 was not quite as good as that of the Cerac material. The poor low-temperature performance of the synthetic CoS_2 mirrors that observed for the synthetic FeS_2 (Figure 3a) and appears to be associated with the nanostructured nature of the disulfides. The difference in performance of these materials at 500°C (Figure 4b) was much less than at 400°C. The USN CoS_2 heat treated at 550°C performed as good the Cerac CoS_2. The materials heat treated at 400° and 500°C shows the same voltage transition near 0.7 eq. Li/CoS_2 as they did at 400°C but with an additional voltage plateau starting near 1.35 eq. Li/CoS_2. The high-temperature performance is consistent with the TGA data for these materials, with materials showing a number of weight-loss transitions prior to the major decomposition reaction also exhibiting additional voltage

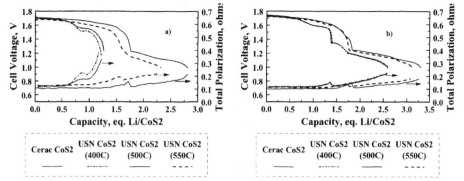

Figure 4. Performance of Synthetic CoS$_2$ in Li(Si)/LiBr-KBr-LiCl (MgO)/CoS$_2$ Cells at 125 mA/cm^2 at a) 400°C and b) 500°C for Various Heat Treatments.

plateaus. The electrochemical signatures of these materials are more sensitive for materials characterization than either XRD or TGA.

5-Cell Battery Tests – Five-cell batteries were built with the USN FeS$_2$ and CoS$_2$ for comparison to control batteries built with natural pyrite and Cerac CoS$_2$. These were tested at an activation temperature of 74°C. The FeS$_2$-based batteries used a heat balance of 95.3 cal/g of total cell mass, while the CoS$_2$ analogues used a heat balance of 96.9 cal/g of total cell mass. (This is the amount of heat needed to raise the stack temperature to operating temperature.) The performance of the FeS$_2$ batteries is summarized in Figure 5a and the corresponding data for the CoS$_2$ batteries are presented in Figure 5b. (The CoS$_2$ from USN had been heat treated at 500°C.)

The battery performance is directly affected by the stack temperature. In the case of the FeS$_2$ batteries, the stack temperatures for the two batteries were almost identical with very comparable electrical performance (Figure 5a). The battery data corroborate the single-cell performance at 500°C (Figure 3b). If the batteries had run longer, then the effects noted for the performance of single cells at 400°C for the synthetic pyrite would likely have also been observed for the cooled battery cooled if more capacity had been removed as it was for the single-cell tests.

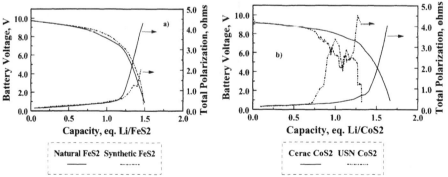

Figure 5. Performance of Synthetic Disulfides in Li(Si)/LiBr-KBr-LiCl (MgO)/MS$_2$ 5-Cell Batteries Activated at 74°C at 125 mA/cm^2. a) FeS$_2$ and b) CoS$_2$.

211

In the case of the CoS_2 batteries, the overall performance of the unit with the USN CoS_2 treated at 500°C was poor compared to that of the battery with Cerac CoS_2 (Figure 5b). In this case, the battery data were corroborated with the poor single-cell performance observed at both 400° and 500°C (Figures 4a and 4b). This was not due to temperature effects, since the thermal profiles were quite similar, but must result from the intrinsic properties of the CoS_2 that cause difficulties during normal catholyte processing. The composition of electrolyte and processing temperatures used in catholyte processing influence interparticle contact and particle wetting by electrolyte. Experiments are underway to improved particle wetting by electrolyte by several pretreatment procedures. This should then lead to improvements in the electrochemical performance. Similar positive effects may also be realized by increasing the CoS_2 particle size.

CONCLUSIONS

The preparation of phase-pure nanostructured FeS_2 and CoS_2 by an aqueous process was demonstrated. These are amorphous as prepared but become crystalline upon heat treatment at temperatures up to 400°C for FeS_2 and 550°C for CoS_2. The thermal treatment has a major impact on the final thermal stability and phase purity of the disulfides, with the best overall properties being observed at the higher treatment temperatures. When tested in single cells at 500°C and 125 mA/cm^2, the USN FeS_2 performs as well as natural pyrite. However, poor overall performance is obtained at 400°C. Similar results are obtained under the same discharge conditions with the USN CoS_2 heat treated at 550°C but not as severe as for the FeS_2. Good performance of the USN FeS_2 is observed in a 5-cell battery activated at 74°C and discharged under the same conditions. However, the corresponding performance of a 5-cell battery made with 500°C USN CoS_2 is unacceptable. These differences are not temperature related but have to do with the nanostructure and morphology of the disulfides and how they are heat treated and processed into catholyte. The high impedance observed during discharge may be related to inadequate electrolyte wetting of the first discharge phase or interfacial problems. These inadequacies are being addressed by pretreatment of the disulfides and possibly by increasing the particle size. These materials still show good potential for use in thermal batteries.

REFERENCES

1. S.V. Kozerenko, D. A. Khramov, V. V. Fadeev, A. M Kalinichenko, I. N. Marov, G. A. Evtikova, and V. S. Rusakov, *Gheokhimiya,* **9**, 1352 (1995).
2. G. Ronsenthal, *Heidelb. Beltr. Mineral. Petrogr.,* **5**, 146 (1956).
3. D. T. Rickard, *Amer. J. of Science,* **275**, 636 (1975).
4. G. W. Luther, III, *Geochim. et Cosmochim.,* **55**, 2839 (1991).

ACKNOWLEDGMENTS

Sandia National Laboratories is a multiprogram laboratory operated by Sandia Corp., a Lockheed Martin company, for the United States Department of Energy under Contract DE-AC04-94AL85000.

This work was performed as a joint effort under a Phase II SBIR contract from the U.S. Army (DAAH01-98-C-R046).

Mat. Res. Soc. Symp. Proc. Vol. 730 © 2002 Materials Research Society V7.4

Charge density in disordered boron carbide:$B_{12}C_3$. An experimental and ab-initio study.

Gianguido Baldinozzi, Michaël Dutheil, David Simeone[1] and Andreas Leithe-Jasper[2]
SPMS, CNRS Ecole Centrale Paris, F 92295 Châtenay-Malabry, France
[1]LM2E, CEA, CE Saclay, F 91191 Gif-sur-Yvette, France
[2]NIRIM, Namiki 1-1, Tsukuba Ibaraki 305-0044, Japan

ABSTRACT

A charge density study of boron carbide $B_{12}C_3$ single crystals at different temperatures allows a quantitative description of the electron density responsible for chemical bonding. These results, based on direct observations, are compared to previous models of bonding. This description points out that structural features are more complex than suggested by simple qualitative models. The effects of chemical substitution, the characteristics of the interatomic bonding, the charge transfer and the bond strength are discussed and compared to the information obtained from previous conventional refinements and ab-initio calculations in disordered $B_{12}C_3$.

INTRODUCTION

Among the members of the boron carbide family the most interesting and perhaps more studied compound is $B_{12}C_3$, which possesses remarkable properties: the melting temperature is very high and the mechanical properties (hard as corundum, stronger than steel, lighter than aluminium) are outstanding [1]. These properties are related to the unusual bonds occurring between boron and carbon atoms [2]. The average structure of $B_{12}C_3$ is trigonal R-3m. This structure consists of a molecular-like sub-unit based on 12 atoms forming an icosahedron sitting at the origin of the primitive trigonal cell and a linear chain extending along the threefold axis (Fig. 1).

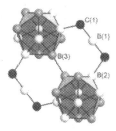

Figure 1. Structure of boron carbide.

The existence of a threefold axis and of an inversion centre at the origin requires two independent atoms, B(2) and B(3), to describe the icosahedra. It is therefore possible to distinguish between 6 equatorial sites B(2) (forming bonds with the linear chain) and 6 polar sites B(3) (forming bonds with the neighbouring icosahedra). The actual location of carbon substitution for boron is still an open issue though several clues [3-16] suggest a C-B-C chain and a C atom somewhere on an icosahedral site. Our aim is to elucidate these bonding features through an experimental study of the charge density on a single crystal of $B_{12}C_3$. All the previous structural determinations of $B_{12}C_3$ were performed on powders or polycrystalline samples. Here, we present various structural refinements on a single crystal using different models of increasing complexity and discussing the validity of the approximations. We start from a standard structure refinement using spherical atomic factors, then we discuss the static and dynamic disorder in the structure before refining the structure within the Hansen-Coppens multipolar formalism [17] for the atomic scattering factors. The results obtained within this approximation are then interpreted and the topology of the chemical bonding is compared to the results obtained by ab-initio modelling.

EXPERIMENTAL

High purity isotopic single crystals of $^{11}B_{12}C_3$ were grown for the first time using an IR image furnace by NEC Machinery Corporation. Heating is achieved by focusing the radiation from a Xe lamp using elliptical mirrors. The obtained single crystals are cylinders regularly shaped (10 mm diameter, several mm height) growing along the [101] direction. After cutting small parts of the large single crystals, several small regularly shaped crystals (prisms of about 100 µm side) were selected and checked for performing the x-ray data collection. The diffraction data were collected on a standard Brucker Smart/ccd diffractometer equipped with an Oxford cryostream. Two sets of data were collected at room temperature and at 100 K using monochromatic MoKα radiation (0.71073 Å). The unit cell parameters are in extremely good agreement with the previous determinations on stoichiometric powder samples of $B_{12}C_3$ [3]. The complete Ewald sphere was scanned leading to about 7500 recorded reflections (462 unique). No absorption and extinction corrections were necessary.

RESULTS

The standard refinement of the structure gives a result very close to the one of $B_{12}C_3$ already studied by Kirfel et al [18]. We have therefore strong experimental evidence that in $B_{12}C_3$ the atom sitting in the middle of the linear chain is a B atom [19]. Nevertheless, in this structure, only two C atoms were located at this time, both sitting at the ends of the linear chain. To locate the remaining C atom, we had a closer look at the spherical model including anisotropic harmonic thermal displacement parameters for all atoms. The refinements of the atomic positions and of the thermal displacement parameters at room temperature and at 100 K are satisfactory for all atoms but the one sitting in the middle of the linear chain. This atom is characterised by a prolate thermal ellipsoid at both temperatures, suggesting a static character of the thermal displacement in the direction along the threefold axis. To provide a better description for this effect, anharmonic thermal displacement parameters for B(1) were refined within the Gram-Charlier approximation [20]. The lowest terms allowed by symmetry, beyond the second order harmonic ones, are fourth order and the improvement of the reliability factors is significant. The obtained one-particle potential for the B(1) atom is displayed on Fig. 2. This potential clearly exhibits a weak anharmonic shape characterised by two shallow minima along the threefold axis at 0.07 Å from the ideal central position; the thermal vibrations, even at 100 K, are sufficient to average the occupancies of the two wells, suggesting a dynamic nature for this disorder. An explanation for the shape of the anharmonic potential comes from the possible bonds formed by B(1) atoms. B atom valence shells contain three electrons.

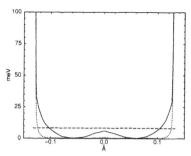

Figure 2. One particle potential for B(1) at 100 K. The horizontal dashed line represents the thermal averaging. The dotted line is the error estimated by Monte-Carlo [20].

The experimental result suggests each B(1) atom forms a double bond with one of the C(1) atoms of the linear chain and a simple bond with the opposite one, and B(1) sits in one of the two minima. At this point it would be very seducing to try to apply the same analysis of the thermal displacements to the icosahedral sites and to try to find out a signature for locating the remaining C atom. The third C atom should sit somewhere on one of the icosahedral sites. If this substitution were not random, the true symmetry of the structure would be monoclinic. A careful analysis of the experimental data does not support this symmetry and a trigonal space group was definitely taken for modelling $B_{12}C_3$. Two possibilities, not implying a symmetry lowering, should also be analysed: either a random selective substitution on the polar (or equatorial) sites or a totally random substitution on the 12 sites of the icosahedron. Unfortunately, the lower symmetry of the B(2) and B(3) sites and the random distribution of the C atom on at least 6 positions reduce the sensitivity of a direct refinement of the occupancies and of the thermal displacement parameters. A more sophisticated way to measure the impact of the C substitution on the icosahedral sites is the multipolar approach [17]. The splitting of the core and valence shell contributions to the diffracted intensities is obtained using the usual Hansen-Coppens formalism by a nucleus-centred multipole expansion. The multipole coefficients of the atoms were constrained to obey the point symmetries and we kept the significant multipoles during the refinements. A global cell electroneutrality constraint corresponding to the $B_{12}C_3$ stoichiometry was used in all refinements. The refinements were performed using the Xd program [21]. Within this model, it clearly appears that the B(3) atom has a mean valence charge larger than the expected 3 electrons. Therefore the C substitution is selective and it occurs on the polar sites. A further confirmation was obtained by Hartree-Fock calculations with Crystal95 [22] on the structure of $B_{12}C_3$ using a simple STO-6G basis set to obtain a reference for the static electron density.

Table I. Valence population (P_v) and shell contraction (κ); δ is the negative charge of the chain.

	Experiment				Ab-initio $B_{12}C_3$		Ab-initio $B_{13}C_2$	
	x=-y	z	P_v	κ	P_v	κ	P_v	κ
C(1)	-1/3	-0.0480(1)	4.28(1)	0.949(2)	4.46(1)	0.954(5)	4.23(1)	0.949(4)
B(1)	-1/3	-1/6	2.57(1)	1.013(1)	2.27(1)	1.099(1)	2.63(1)	1.031(7)
B(2)	-0.1703(1)	-0.0250(1)	2.98(1)	1.034(4)	3.02(1)	1.014(4)	2.99(1)	1.006(3)
B(3)	-0.1074(1)	0.1138(1)	3.34(1)	1.021(3)	3.18(1)	1.072(3)	2.97(1)	1.003(3)
δ			0.13(1)		0.19(1)		0.09(1)	

We have calculated the charge density and computed the x-ray structure factors for two different compounds: stoichiometric $B_{13}C_2$ and C enriched $B_{12}C_3$. In the latter compound, an ordered selective C substitution on one of the B(3) sites was assumed. Therefore, the full sphere of reflections was generated, merged according to the mean R-3m space group to restore the average trigonal symmetry and cut at the same experimental resolution. These calculations confirm the effect of the selective C substitution observed in the experiment. The quality of the experimental refinement is assessed by the good agreement factor (2.5 %), compared to 1.5 % obtained on the $B_{12}C_3$ ab-initio structure and 0.9 % for ab-initio $B_{13}C_2$. The charge transfer between the chain and the icosahedron (Tab. I) determined from our experimental model is 0.16 e, slightly smaller than the value obtained from the ab-initio calculation of B_{11}C-CBC. This is not surprising since in the calculated model B(1) is sitting exactly in the middle of the chain. It is therefore more instructive to compare our results with the experimental value of 0.07 e in $B_{13}C_2$ [18]. The increase of charge transfer with C concentration is in agreement with the trend suggested by ab-initio calculations [9,15,16]. Sections of the static multipolar deformation densities obtained for experimental $B_{12}C_3$, ab-initio $B_{12}C_3$ and ab-initio $B_{13}C_2$ in the plane containing the five independent atoms are shown in Fig. 3. The comparison between these models outlines the following features:

- the C-B-C chain is well described in each model, which confirms that we correctly took into account the positional disorder affecting B(1),
- the substitution of C for B occurring on B(3) strengthens the average inter-icosahedral bond,
- the intra-icosahedral bonds are largely affected by the C substitution in the experimental model,
- the charge depletion at the icosahedron centre is increased in the C enriched compounds and seems to be associated with a relaxation of the bonds.

To obtain a more quantitative description of the bonds in $B_{12}C_3$, a topological analysis of the static deformation density was performed with Xd and the different critical points of the interatomic bonds in this compound were analysed. To estimate the accuracy of the Laplacian of our experimental data, this was also computed for the ab-initio electronic densities. Bond type, position of the critical points and a list of the density properties describing the relative strength of the most interesting chemical bonds are summarised in Tab. II. The existence of important negative values in the eigenvalues of the Hessian and the large value of the ellipticity of some of them clearly suggests a π character for the covalent intra and inter-icosahedral bonds.

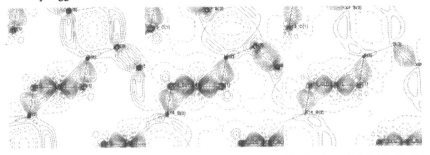

Figure 3. Static deformation densities for the experiment, ab-initio $B_{12}C_3$ and ab-initio $B_{13}C_2$.

Table II. Topological properties of the static multipolar densities. The intericosahedral bonds are marked with *. ρ is the charge density at the critical point ($e\text{Å}^{-3}$), $\Delta\rho$ the Laplacian, $\lambda_{1,2}$ the curvatures normal to the bond, λ_3 along the bond, ε the bond ellipticity, G, V and E the local kinetic, potential and total energy respectively [21].

Experiment	ρ	$\Delta\rho$	λ_1	λ_2	λ_3	ε	G	V	E
B(1)-C(1)	1.65(1)	-17.06(1)	-10.94	-10.94	4.83	0.00	3.75	-11.77	-8.02
C(1)-B(2)	1.18(2)	-10.50(5)	-6.73	-6.71	2.94	0.00	2.02	-6.67	-4.65
B(3)-B(3)	0.89(1)	-2.89(2)	-4.61	-1.46	3.18	2.16	1.88	-4.49	-2.60
B(3)-B(3)/C(3)*	0.87(1)	-2.31(3)	-3.31	-2.73	3.73	0.21	1.89	-4.37	-2.47
B(2)-B(2)	0.83(1)	-2.36(2)	-3.79	-1.74	3.17	1.18	1.71	-4.01	-2.30
B(2)-B(3)	0.74(1)	-0.95(2)	-3.31	-0.74	3.11	3.47	1.57	-3.39	-1.81
Ab-initio $B_{12}C_3$									
B(1)-C(1)	1.729(3)	-16.12(1)	-12.24	-12.24	8.36	0.00	4.47	-12.96	-8.49
C(1)-B(2)	1.194(6)	-9.65(3)	-8.39	-7.71	6.45	0.09	2.25	-6.91	-4.66
B(3)-B(3)	0.739(4)	-0.45(1)	-3.23	-1.18	3.96	1.73	1.66	-3.43	-1.77
B(3)-B(3)/C(3)*	1.093(7)	-5.81(2)	-5.33	-5.16	4.68	0.03	2.36	-6.18	-3.81
B(2)-B(2)	0.859(5)	-2.86(1)	-3.81	-1.83	2.78	1.08	1.75	-4.22	-2.47
B(2)-B(3)	0.781(4)	-1.43(1)	-3.57	-1.08	3.22	2.30	1.66	-3.68	-2.02
Ab-initio $B_{13}C_2$									
B(1)-C(1)	1.475(3)	-6.25(1)	-10.10	-10.10	13.95	0.00	4.45	-10.45	-6.01
C(1)-B(2)	1.078(8)	-2.04(3)	-7.04	-6.02	11.02	0.17	2.91	-6.34	-3.42
B(3)-B(3)	0.734(5)	-0.88(1)	-2.67	-0.93	2.72	1.88	1.57	-3.36	-1.79
B(3)-B(3)*	0.982(5)	-7.90(1)	-5.06	-4.91	2.12	0.03	1.48	-4.92	-3.44
B(2)-B(2)	0.817(4)	-2.69(1)	-3.36	-1.81	2.49	0.86	1.60	-3.88	-2.27
B(2)-B(3)	0.732(5)	-1.20(1)	-2.89	-0.95	2.64	2.06	1.51	-3.31	-1.81

Following energetic arguments, the inter-icosahedral B(3)-B(3) polar bond is the strongest bond in the structure after B(1)-C(1) both in B_{11}C-CBC and in B_{12}-CBC ab-initio structures. In our experimental charge density on B_{11}C-CBC, the polar and the inter-icosahedral bonds have strengths in agreement with the ab-initio results within the limit of the estimated errors. The different local densities at the critical points for the icosahedral bonds arise from the fact that only an average atom B(3) was defined in the multipolar refinement. Therefore, the effect of averaging and the effect of substitution are almost as important, but these two effects cannot be split in the experiment. The Laplacian of the density presents a behaviour less regular then the charge density. These discrepancies are more pronounced for the B(1) atom since the multipolar development for this atom is centred in the middle of the chain, representing only the average behaviour of the real bonds formed with C(1). Therefore, the trick of handling the positional disorder within a Gram-Charlier formalism is a good approximation for the charge density. The analysis of the experiment shows that the C substitution on the polar sites decreases the strength of the substituted inter-icosahedral bond B(3)-B(3). This effect is confirmed by accurate ab-initio calculations [9,15,16] in the primitive monoclinic cell. Nevertheless, inter-icosahedral bonds are still the strongest in the structure (and stronger than in $B_{13}C_2$), after those involving the atoms in the CBC chains. Therefore, the icosahedra behave like anti-molecular units.

CONCLUSION

Our study was mainly concerned with the qualitative description of the bond properties of disordered $B_{12}C_3$ compound. These results are in good agreement with the predictions of ab-initio calculations, but there are still some inaccuracies in the model of the experiment, possibly related to the limits of the multipolar formalism. The different kinds of disorder affecting this structure represent a limitation for the study of the topology of the charge density. To circumvent this problem and to obtain a reliable model for understanding the complex features and the bonding in this complex structure is coupling experiment and ab-initio calculations. The analysis of these results gives direct evidence for the random substitution of one carbon atom onto one of the six polar sites in the icosahedron and for the effects of this substitution, leading in particular to an average strengthening of the inter-icosahedral bonds. The resulting picture could be a useful starting point for a re-analysis of the conductivity mechanisms in this family of compounds.

REFERENCES

1. D. Emin, *Phys. Today* **20, 55** (1987).
2. H. C. Longuet-Higgins and M. Roberts, *Proc. R. Soc. London A* **230**, 110 (1955).
3. M. Bouchacourt and F. Thevenot, *J. of Less Common Metals* **82**, 227 (1981), 227.
4. B. Morosin, A. W. Mullendore and D. Emin, G. A. Slack, *AIP Conf Proc* **140**, 70 (1987).
5. B. Morosin, G. H. Kwei, A. C. Lawson, T. L. Aselage and D. Emin, *J. Alloys and Compounds* **226**, 121 (1995).
6. D. Tallant, T. Aselage, A. Campbell and D. Emin, *Phys. Rev. B* **40**, 5649 (1989).
7. A.Howard, C. L. Beckel and D. Emin, *Phys. Rev. B* **35**, 9265 (1987).
8. D. M. Bylander, L. Kleinman and S. Lee, *Phys. Rev. B* **42**, 1394 (1990).
9. D. Li and W. Y. Ching, *Phys. Rev. B* **52**, 17073 (1995).
10. D. Simeone, C. Mallet, P. Dubuisson, G. Baldinozzi, C. Gervais and J. Maquet, *J. Nucl. Mat.* **277**, 1 (2000).
11. H. H. Madden, G. C. Nelson, *Phys. Rev B* **31**, 3667 (1985).
12. D. Emin, *Phys. Rev B* **38**, 6041 (1988).
13. T. Harazono, Y. Hiroyama and T. Watanabe, Bull. Chem. Soc. Jpn. **69**, 2419 (1996)
14. H. Yackel, *Acta Cryst.* B **31**, 1797 (1975).
15. R. Lazzari, N. Vast, J. M. Besson, S. Baroni and A. dal Corso, *Phys. Rev. Lett.* **83** , 3230 (1999).
16. N. Vast, J. M. Besson, S. Baroni and A. Dal Corso, *Comp. Mat. Sc.* **17**, 127 (2000).
17. N. K. Hansen and P. Coppens, *Acta Crystallogr.* A **34** , 909 (1978).
18. A.Kirfel, A. Gupta and G. Will, *Acta Crystallogr.* B **35** 1052 (1979), B **35** 2291 (1979), B **36**, 1311 (1980).
19. A.Silver and P. Bray, *J. of Chem. Phys.* **31**, 247 (1959).
20. W. F. Kuhs, *Acta Crystallogr.* A **48**, 80 (1992).
21. T. Koritsanszky, S. Howard, T. Richter, Z. Su, P.R. Mallinson and N.K. Hansen, *Xd - A Program Package for Multipole Refinement and Analysis of Electron Densities from Diffraction Data*, Free University, Berlin, (1995).
22. R. Dovesi, V. R. Saunders, C. Roetti, M. Causa, N. M. Harrison, R. Orlando and E. Apra, *Crystal95 User Manual*, University of Torino, Torino (1995).

Thermoelectrics

Mat. Res. Soc. Symp. Proc. Vol. 730 © 2002 Materials Research Society　　　　　　　　　　　V8.1

Anomaly of Thermal Properties in Thin Films of La$_{1-x}$Sr$_x$CoO$_3$ Series Fabricated as Thermoelectric Materials

Yoshiaki Takata[1,2], Hajime Haneda[1,2], Yutaka Adachi[2], Yoshiki Wada[2], Takefumi Mitsuhashi[2], Makoto Ohtani[3], Tomoteru Fukumura[1,4], Masashi Kawasaki[1,4], and Hideomi Koinuma[1,5]
[1]Combinatorial Materials Exploration and Technology (COMET),
Namiki, Tsukuba, 3050044, Japan.
[2]Advanced Materials Laboratory, National Institute for Materials Science,
Namiki, Tsukuba, 3050044, Japan.
[3]Innovative and Engineered Materials, Tokyo Institute of Technology,
Midori-ku, Yokohama, 2268503, Japan.
[4]Institute for Materials Research, Tohoku University,
Aoba-ku, Sendai, 9808577, Japan.
[5]Frontier Collaborative Research Center, Tokyo Institute of Technology,
Midori-ku, Yokohama, 2268503, Japan.

ABSTRACT

Composition spreads of La$_{1-x}$Sr$_x$CoO$_3$ (LSCO) were synthesized as thermo-electric transducer material by means of combinatorial material synthesis, and information on their thermal diffusivity throughout the specimens was obtained. Meanwhile, it is anticipated that the composition spreads of LSCO have a variety of characteristic properties of light absorption corresponding to their composition, namely their compositional variable, x. Hence, we used transient optical pump-and-probe techniques in a reflection geometry to measure thermal diffusion times on LSCO. The results of signal analysis indicate that LSCO is a peculiar synthesized substance whose apparent thermal diffusivity, $1/\tau$, changes abruptly at the critical point, where the compositional variable x is equal to 0.3.

INTRODUCTION

It is hopefully expected that the invention of composition-spread preparation by synthesizing combinatorial materials helps to accelerate development of new materials [1]. With this background, the authors first synthesized a composition spread of LSCO as material for thermoelectric transducers, and evaluated thermal diffusivity as an important characteristic property [2]. Because the LSCO substances have a variety of characteristic properties of light absorption depending upon the compositional variable, x, we believe that there are no conventional analytical techniques versatile enough to analyze comprehensively the thermal diffusion of these synthesized substances. In this work, we have employed transient optical pump-and-probe techniques to heat the samples and measure transient thermal diffusion times. In these measurements, the third harmonic (266 nm) of a modelocked Ti:sapphire laser was used to heat a small spot on the sample. The heat-induced change in reflectivity was measured by probing the heated spot with light pulses at the fundamental wavelength (800 nm) as a function of time delay between the pump and probe pulses. Hence, this study evaluated the apparent thermal diffusivity of LSCO, by observing the time-resolved reflection of light, and we report on the apparent thermal diffusivity so obtained.

EXPERIMENTAL DETAILS

The following items were employed in our experiment: This study was carried out by observing signals produced from two-pump-one-probe method, in other words so-called compound signals. The respective signals at the measurement points, i.e., the position of the specimen, were scanned and observed at every 0.5 mm interval of space in sequence. The creative preparation of samples of the LSCO series has been described in detail previously [1]. Described briefly, however, two kinds of ceramic targets, $LaCoO_3$ and $SrCoO_3$, are used in this experiment. The material of the substrate is $SrTiO_3$ (100), and the dimensions of the substrate are 14 mm x 7 mm x 0.5 mm in size. A schematic diagram of the sample specimen used for measurement of apparent thermal diffusivity is outlined in Figure 1. The central 5 mm region in this figure represents the composition-spread film comprising one of the LSCO series, the left-side 2 mm region is of single-compound $LaCoO_3$ film, while the right-side 2 mm region is of single compound $SrCoO_3$ film. The thickness of each film is 150 nm. Both ends are the substrate material, $SrTiO_3$.

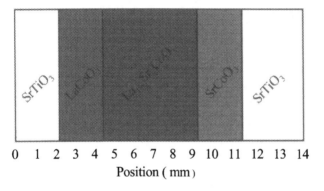

Figure 1. A schematic diagram of the sample specimen used for measurement of apparent thermal diffusivity.

In our measurements, we have used a regeneratively-amplified Ti:sapphire laser system in which the Ti:sapphire oscillator (Mira900) is pumped by a 5 W cw-Nd:YVO$_4$ laser and the Ti:sapphire regenerative amplifier (RegA9000) is pumped by an Ar ion gas laser (Sabre, 14 W, made by Coherent). The laser-light emitted from the RegA9000 has wavelength of 800 nm, pulse-width of 200 femtoseconds, repetition frequency of 200 kHz, and output power of 1 W. The laser-light produced as stated above passes through a TP-1B fs optical frequency tripler made by U-Oplaz Technologies, and light pulses at three wavelengths (the fundamental wave at 800 nm, the second harmonic at 400 nm, and the third harmonic at 266 nm, respectively) emerge at the exit of the light path. Among them, the third harmonic pulses whose output power is 50 mW are used for pumping, and the fundamental pulses for probing after they have passed through an optical filter, O-59. The intersecting angle of two pumping pulses is kept at 13° on the surface of the specimen, and the two pumping pulses hold the same light path in common. The geometry of the spot on the specimen-surface is $100 \cdot 200$ (μm^2), and the entering angle of the probe-pulse is 2° off of normal incidence on the same side as pump pulse A. An optical filter,

R72, is attached to the probe-pulse detector so as to prevent light from the pumping pulse from entering and affecting the signals.

SIGNAL

The true characteristic curves of the signals resulting from the transient optical reflectivity measurements are shown in Figure 2. As seen in the three curves, (a), (b), and (c), because there are no specific waves of certain amplitudes on these curves, the signals apparently do not include any interference fringes produced by use of a pair of pump pulses. The curve, (a), represents signals produced by the two-pump-one-probe method, namely so-called compound signals, whereas the curves, (b) and (c), signals by one-pump-one-probe method, namely those produced when either one of the two pump pulses is blocked, or so-called single signal. As shown with the curves, (b) and (c), the signals still appear, albeit their intensities are reduced, in spite of one of the two pump pulses being blocked. We also note that the sum of the signals of (b) and (c) is equal to that of (a).

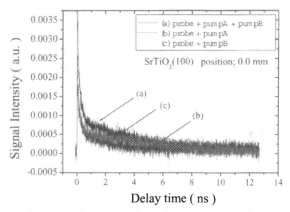

Figure 2. The true characteristic curves of the signals resulting from the transient reflection measurements. (a); the curve by two-pump-one-probe method. (b) and (c); curves by one-pump-one-probe method.

RESULTS AND DISCUSSION

Typical curves of obtained signals are shown in Figure 3. With regard to (a), this figure represents the characteristic curve of a substrate SrTiO$_3$ (100), and (b); that of a simple, pure film comprising SrCoO$_3$. Circles in these figures stand for the values as they were just really measured, whereas the solid lines for theoretically calculated values. Then the method of nonlinear least squares was applied for curve-fitting by using A. Harata and others' [3] empirical equation (1) so as to extract the thermal decay constant, τ, out of each curve. More specifically, the curve-fitting operation was applied for each curve except the part for the first 0.5 ns that was deleted from our calculation.

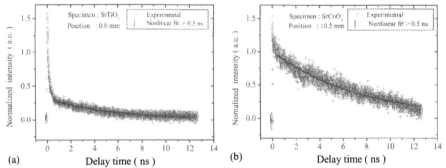

Figure 3. The typical curves of both the experimental signal and the calculated curve when the measurement positions are (a) SrTiO$_3$(100) at 0.0 mm and (b) SrCoO$_3$ at 10.5 mm in the LSCO.

$$S(t) = A\{\exp(-t/\tau) - r\exp(-t/\tau_w)\cos[2\pi f_w(t + t_w)]\}^2 \qquad (1)$$

where, τ stands for thermal decay time, τ_w decay time of the surface wave; f_w frequency of the surface wave; t_w the decay time in the surface wave generation; r the coupling ratio of the thermal decay element and surface wave.

Results of curve-fitting and the respectively calculated thermal decay constants.

Now, the thermal decay times τ are: 11 ns for (a), and 15 ns for (b). Hence, we hesitated here to calculate thermal diffusivity by using these obtained thermal constants, τ, taking into consideration that the obtained data on the signals might not result from the Bragg diffraction of light, but may be from scattered reflection light. Instead, we carried out our operation,

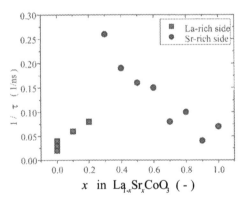

Figure 4. Apparent thermal diffusivity, $1/\tau$, the composition variable (x) in the LSCO, and relationship with x of all the samples.

empirically assuming that $1/\tau$ would represent apparent thermal diffusivity. This apparent thermal diffusivity, $1/\tau$, the composition variable (x) in the material, and relationship with x of all the samples are shown in Figure 4, wherein the data represented by squares indicates a region where the material is considered to be La-rich, while the data represented by circles a region where the material is considered to be Sr-rich.

As shown in Figure 4, the maximum value of $1/\tau$ appears at $x = 0.3$, and in addition to it, at this critical point the curve shows a peculiar point of inflection (a fault-like profile) at $x = 0.3$ on this curve, which corresponds to the position, 0.6 mm, as returned to the measurement point. The abrupt change of the thermal characteristics before and after this point seems to be approximately that reported in our previous paper [2]. Furthermore, these apparent thermal diffusivity, $1/\tau$, the intensity of X-ray diffraction (XRD), and relationship with intensity of XRD of all the samples are shown in Figure 5, wherein the data represented by squares indicates a region where the material is considered to be La-rich, while that represented by circles a region where the material to be Sr-rich. Circles and squares in this figure stand for the measured values, whereas the solid lines for linearly fitted values. The linear-fittings were applied to classify the difference of characteristic properties between La-rich and Sr-rich compound. As shown in Figure 5, a larger increase of the apparent thermal diffusivity, $1/\tau$, appears in Sr-rich compounds than La-rich. The different changes of the thermal characteristics between Sr-rich and La-rich compounds seem to be approximately those reported in our previous paper, too [2]. It is suggested, in other words, that the thermal diffusion times extracted from measurements of Bragg diffraction of light seem very similar to those resulting from the scattering reflection of light in the present paper.

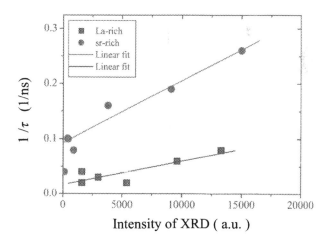

Figure 5. The relationship between the intensity of X-ray diffraction, and the apparent thermal diffusivity, $1/\tau$, throughout the sample specimen.

CONCLUSIONS

Transient optical pump-and-probe techniques, in which the third harmonic from Ti-sapphire laser light is employed as a pump source to heat the sample, enable us to recognize the scattering reflection light that gives information on the thermal diffusion in accordance with the compositional variable, x, in LSCO. However, the problem yet to be solved is to convert apparent thermal diffusivity, $1/\tau$, into numerical results for thermal diffusivity.

REFERENCES

1. T. Fukumura, Y. Okimoto, M. Ohtani, T. Kageyama, T. Koida, M. Kawasaki, T. Hasegawa, Y. Tokura and H. Koinuma, *Appl. Phys. Lett.* **77**, 3426 (2000).
2. Y. Takata, Y. Adachi, H. Haneda, Y. Wada, T. Mitsuhashi, M. Ohtani, T. Fukumura, M. Kawasaki and H. Koinuma, in *Combinatorial and Artificial Intelligence Methods in Materials Science*, edited by I. Takeuchi, J. M. Newsam, L. T. Wille, H. Koinuma and E. J. Amis, (Mater. Res. Soc. Proc. **700**, Boston, MA, 2001) pp. 167-172.
3. A. Harata, H. Nishimura and T. Sawada, *Appl. Phys. Lett.* **57**, 132-134 (1990).

AUTHOR INDEX

SUBJECT INDEX

Printed in the United States
By Bookmasters